AMERICAN ENTERPRISE

National Museum *of* American History

DAVID K. ALLISON | NANCY DAVIS | KATHLEEN G. FRANZ | PETER LIEBHOLD

AMERICAN ENTERPRISE

A History *of* Business in America

contributors edited *by* **ANDY SERWER**

SHEILA BAIR

ADAM DAVIDSON

BILL FORD

SALLY GREENBERG

FISK JOHNSON

HENRY M. PAULSON JR.

RICHARD L. TRUMKA

PATRICIA WOERTZ

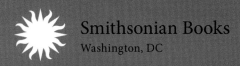

Smithsonian Books
Washington, DC

This book may be purchased for educational, business, or sales promotional use. For information, please write:
Special Markets Department
Smithsonian Books
P. O. Box 37012
MRC 513
Washington, DC 20013

Published by Smithsonian Books
Director: Carolyn Gleason
Production Editor: Christina Wiginton
Editorial Assistant: Raychel Rapazza
Museum Coordinator: Katharine Klein

Edited by Evie Righter
Designed by Bill Anton

Library of Congress Cataloging-in-Publication Data
American enterprise: a history of business in America / edited by Andy Serwer; introduction by David K. Allison; contributions by Peter Liebhold; contributions by Nancy Davis.
    pages cm
ISBN 978-1-58834-496-0 (hardback)
1. United States—Commerce—History. 2. National characteristics, American 3. United States—Economic conditions. 4. Democracy—United States—History.
5. Capitalism—United States—History. I. Serwer, Andy.
II. National Museum of American History (U.S.)

HF3021.A44 2015

338.0973—dc23   2014027166

Manufactured in Canada
19 18 17 16 15   5 4 3 2 1

For permission to reproduce illustrations appearing in this book, please correspond directly with the owners of the works, as seen on p. 208-210. Smithsonian Books does not retain reproduction rights for these images individually, or maintain a file of addresses for sources

*title page: **Patrick Lyon at the Forge**, by John Neagle, 1826–1827.*
After refitting the doors on a bank vault, blacksmith Lyon was falsely convicted of robbing the bank. Imprisoned, he won his acquittal, plus restitution, which he used to manufacture fire engines. Lyon commissioned this portrait to indicate his pride as an American laborer.

Table of Contents

Foreword

WE ARE PLEASED to be presenting the sweep of American Enterprise explored in this publication with its accompanying exhibition and multifaceted programs. Understanding American capitalism is essential to knowing how we as a Nation came to this moment, and even more importantly, what our future could be. As the museum's first major initiative to present American business to the public, we take this very seriously, and are fascinated by the stories of such varied American innovators.

My favorite story in this book is about Tupperware. In the 1940s, Earl Tupper, an independent inventor and entrepreneur, determined an effective way to mold plastic into safe food containers and—here's the key—include a top that made an airtight seal. His Wonder Bowls worked well, but Tupper failed to find effective ways to sell them. So he developed an unlikely partnership with a woman named Brownie Wise, who had begun marketing Tupper's plastic containers through home sales.

Tupper contacted her after learning about her great success with this approach. American consumers, mostly women, were skeptical about plastic and rarely saw the value of the vacuum seal without a demonstration. Although Wise didn't invent the home selling strategy, she adapted it and tapped into a growing desire among married, middle-class women to contribute to the family income. Wise made selling fun and lucrative, and thus turned Tupperware into a household standard.

Today the National Museum of American History preserves elements of this quintessential American story in both dozens of examples of early Tupperware and archival papers from both Tupper and Wise. It is but one example of how the institution is preserving, interpreting, and sharing the remarkable history of *American Enterprise*.

Traditionally our museum presented highlights of its broad collections, which number over three million artifacts and records, and document everything from medicine to music. But in recent years we have learned that our visitors are more interested in seeking answers to provocative questions: what does it mean to be an American? What are the ideas, ideals, and realities that have defined the American experience?

American Enterprise aspires to answer these questions about the history of American business and innovation. It tells the story of what happened when capitalism and democracy met in North America and shaped a nation of people eager for economic opportunity and willing to embrace innovation and change. But this new exhibition will not stand alone. The museum is also exploring related areas, including how democracy was implemented in the United States; how immigration and migration changed national beliefs and values; and how new cultural expressions have shaped American identity at the individual, community, and national level.

Interspersed throughout this book are essays of leading thinkers in today's business community, who present contrasting views on important contemporary issues. Debating enterprise has always been part of American history, from the time when some colonists argued that remaining British was best for business, while others passionately sought independence. I hope this mix of historical narrative and contemporary deliberation will challenge you to define your own views about the best direction for American business and innovation in the future.

JOHN L. GRAY
Director, National Museum of American History

Tupperware pie wedge holder, 1947–1960. Inventor Earl S. Tupper molded his idea for plastic containers into a variety of specialized shapes that could hold any food item, from salad to dessert, and gave consumers many reasons to buy everything in an expanding line of re-sealable products.

Introduction

AUGUST 15, 1786. George Washington, fifty-four years old, recently retired as commanding general of the United States Army, has returned to living as a gentleman farmer at Mount Vernon, his bounteous estate along the banks of the Potomac River. He is writing to his protégé and comrade in arms, the Marquis de Lafayette, now living in France. The topic: how the fledgling United States will manage relations with European powers. Great Britain, Washington believes, is foolishly complacent in its belief that the new nation will remain dependent on it. He says, "However unimportant America may be considered at present, and however Britain may affect to despise her trade, there will assuredly come a day, when this country will have some weight in the scale of empires" (Fitzpatrick 1938, 518–521).

In Washington's eyes, that "weight" should come through expanded commerce, not military power. He continues, "As a citizen of the great republic of humanity at large…I cannot avoid reflecting with pleasure on the probable influence, that commerce may hereafter have on human manners and society in general….I indulge a fond, perhaps an enthusiastic idea, that, as the world is evidently much less barbarous than it has been, its melioration must still be progressive; that nations are becoming more humanized in their policy, that the subjects of ambition and causes for hostility are daily diminishing; and, in fine, that the period is not very remote, when the benefits of a liberal and free commerce will, pretty generally, succeed to the devastations and horrors of war."

Like most of the founders of the United States, Washington was at heart a businessman—not an aristocrat or even a warrior. He had fought for independence primarily for economic, not political reasons. The fundamental question had been who would benefit from the economic growth of America—absentee colonial landlords or the people of America themselves?

The United States was now an independent nation, free to explore and exploit its commercial interests for its own benefit. Just days before Washington wrote Lafayette, the Congress approved an American monetary system based on the "dollar," a name taken from the Spanish silver dollar, or 8 reales coin that circulated widely in the country. Within two centuries, the American dollar would become the most important monetary unit in the world.

Unquestionably, Washington was overly optimistic about the triumph of commerce over war in the interactions among nations. But he was not wrong in discerning that the future of the United States depended on how effectively it balanced its own commercial interests with those of other global powers.

Despite his desire to return to farming, Washington was called back within a year to lead his country, first as President of the Constitutional Convention in Philadelphia, and subsequently as the first President of the United States. In these critical roles, Washington helped establish a nation that would become the world's leading global economy.

Statue of George Washington, 1792. Jean-Antoine Houdon's statue of Washington, begun in 1785, stands in the Virginia State Capitol. Although Washington is depicted in military dress, he is accompanied by both civilian (plow and cane) and military (fasces, sword, and uniform) objects.

United States dollar coin, 1794. In the Coinage Act of 1792, Congress required that the dollar coin have an emblem representing liberty—not the President—on its face, and an eagle on the back. Both became enduring symbols of the nation.

This book, a companion to a major exhibition at the National Museum of American History, surveys the history of the business development of the United States, from the late eighteenth century to the present. The book explores key elements and significant stories that characterize this development. They begin with the location and bounteous natural resources of the United States. The distance of the new nation from Europe allowed it to grow and ultimately expand across the continent without constant interference from the European powers, which continued to be preoccupied with each other. Land, timber, coal, iron ore, and later petroleum among other resources provided the foundation for rapid business development.

Even more important were the people of the new United States. Colonists and then immigrants to the nation came seeking business opportunity, abundant land for farming, and opportunity to work as independent merchants or craftsmen. Slavery brought thousands of Africans to the country against their will and raised the enduring issue of what rights they would have.

Providing the framework for the development of the new nation was its commitment to a pair of overarching ideas of social organization: capitalism and democracy. Capitalism provided an economic system in which private ownership of land and resources would guide development, not state planning and ownership. Under democracy, sovereignty belonged to the people, not a king or a hereditary class of aristocrats. Committed to a sovereign people and free enterprise, the founders of the United States implemented a federal constitution establishing a national government whose powers were limited. It provided for the common defense and regulated interstate commerce and foreign trade. It outlawed aristocracy, protected both physical and intellectual property, and assured individual rights. It established a national monetary system and a postal system to foster the easy flow of information throughout the nation. The government favored private business development rather than state-sponsored enterprise. And it focused on bringing material prosperity to all citizens, not just a class of aristocrats.

There are many ways to divide the history of the United States into distinct periods. One common approach is segmenting it around major military conflicts: the American Revolution, the Mexican War, the Civil War, the World Wars, and Vietnam. Yet, periodizing around wars obscures fundamental changes in business history. Taking a different approach, this book breaks American history into four eras that highlight changing patterns of business organization: a Merchant Era, 1770–1850s; a Corporate Era, 1860s–1930s; a Consumer Era, 1940s–1970s; and a Global Era, 1980s–Present. Certainly these are not fully distinct. For example, small merchants can still flourish today, doing business in ways not much different from those of their forbearers in the early nineteenth century. Likewise, corporations continue to form, innovate, and grow. The intent of this periodization is to highlight new trends that shape economic change.

In the Merchant Era, American businesses and farms were generally small and run by individuals or family members. These included not only blacksmiths, coopers, and shopkeepers, but also most farmers, who not only fed their families, but also grew cash crops, marketed livestock, or sold eggs and milk. Although merchants sometimes had suppliers or customers far afield, most of their business was local, and they would barter goods or time and extend personal credit. Few merchants used banks in daily business.

By the mid-nineteenth century, innovations in water, steam, and later electrical power led to increased industrialization and the growth of larger enterprises, such as mills, textile manufacturers, mines, and transportation companies. These new organizations required large quantities of capital to build and operate, and this need

Leather slippers, about 1840. These leather slippers, handmade in the early nineteenth century, illustrate the skill of American artisans in that era. Soon many would struggle to adjust to the changes brought by industrialization and the loss of control over their work.

Sewing machine room, Brown & Sharpe Manufacturing Company, 1879. Brown & Sharpe, founded in 1833 in Providence, Rhode Island, played a major role in the development of industrial technology. Their facilities for machine tools manufacture illustrate innovative forms of industrial production.

led to new forms of business organization, including trusts, holding companies, and corporations. Key to their expanding role were new laws and regulations that limited liability of investors and stockholders for losses when corporations failed. Although most Americans still worked on farms or in small businesses, corporate growth and regulation by state and federal governments began to dominate economic development.

During this Corporate Era, Americans increasingly bought not locally hand-crafted items, but efficiently mass-produced shoes, clothes, tools, guns, farm equipment, and vehicles. As booming industries produced ever larger numbers of these items, new forms of advertising and marketing developed to sell them to the public.

In the subsequent Consumer Era, successful businesses focused on new forms of consumption. They began targeting segments of the market that would demand customized goods and services. They learned to cater to different racial and ethnic groups, to women as well as men, and to young people as well as adults. New forms of payments such as credit cards were devised to make it easier for consumers to buy what they wanted.

In the 1950s and 1960s, the United States, which had escaped devastation during World War II, became the dominant economic power in the world. Although the nation still had many poor and struggling families, its citizens fared better than those of any other large country. By the 1980s, however, other nations increasingly challenged American dominance. Many adopted techniques copied from American businesses or pioneered innovations that let them produce goods and services

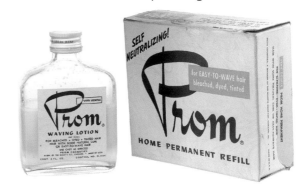

Prom Home Permanent, about 1954. Prom was advertised as a "trouble-free" home permanent product for the "girl who knows beauty." New products like these flooded the American market in the 1950s and 1960s.

at lower cost. Laborers in other nations often worked for lower wages or accepted poorer working conditions. Other nations also adopted trade barriers or laws that restricted sale of American goods.

American business had to become global to succeed, with production as well as sales distributed in locations around the world. Moreover, American consumers demanded access to goods from other countries when they were cheaper or more desirable. Innovations in transportation and information technology, along with the growth of multinational corporations, fostered new forms of international business development and cooperation. Companies that did not adapt to the new global marketplace struggled or failed. As a nation that was created from European colonies, the United States had always been involved in global production, trade, and consumption. But in this era, the depth, pace, and signification of global interaction took on new forms and set the pace for business development.

In surveying all four eras, this book examines business history not only from the perspective of traditional business—the producers of goods and services—but also from the perspective of consumers, and how producers and consumers interact in marketplaces. Over the sweep of history these interactions have ranged from trading hours of labor for a pair of handmade shoes, to shopping in luxurious department stores, to clicking a computer mouse to purchase something with a credit card for overnight delivery. Pacing this interaction between producers and consumers has been the development of advertising and marketing. At times, they simply help consumers learn where they can find products they know they want, but often they also create needs or desires that the consumer never imagined.

Central to business development have been inventors, entrepreneurs, managers, regulators, and other agents of change. Their actions have led to new ideas, technologies, managerial processes, organizations, laws, and procedures that have altered the ways that business is done and products are bought and sold.

Four recurring themes weave through the narrative: opportunity, innovation, competition, and shared concern for common good. The first three of these derive from America's commitment to capitalism and material progress. The fourth relates to the nation's commitment to democracy and a society designed, in the words of the United States Constitution, to "...promote the general Welfare, and secure the Blessings of Liberty to ourselves and our Posterity." All four play out in compelling and dramatic ways in the economic lives of the men and women this book portrays. The outcomes usually have negative as well as positive consequences. Over time, many Americans have benefitted from continual change, but others suffered. Gains in productivity might bring profits to an entrepreneur, but job loss to workers who are no longer needed. Whatever its benefits, American prosperity has not ended widespread poverty, and distribution of business profits has been a contentious issue throughout U.S. history. Nonetheless most Americans have come to believe that continual economic change is inevitable, and that the responsibility of governments and social organizations is to regulate change and ameliorate the negative consequences it brings, not to resist it.

Finally, this book emphasizes material culture as it relates to business history. It is rich with images of objects from the permanent collections of the National Museum of American History. To understand business, it is useful to look closely at the products it produces. This book asks: why do these objects exist? Who made them, and why? What materials were selected and how were the objects produced? What patterns of social interaction do

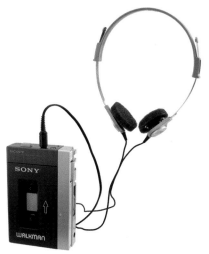

Sony Walkman, about 1980. When Sony introduced its path-breaking Walkman in 1979, it blazed new paths in the American market. Not only did this product pioneer small mobile music players, but it also marked the increasing globalization of the consumer electronics industry.

Underwood #5 typewriter, 1914, and Apple iPhone 4s, 2014. Smart phones combined functions of a host of earlier devices, from typewriters to cameras. Surprisingly, the inefficient QWERTY keyboard persisted.

they represent? How did the interactions among producers, marketers, and consumers shape commodities and everyday consumer goods? How did they accrue value? Were changes readily accepted or resisted? Who gained and who lost when they were produced and sold?

Close attention to material culture in business history reveals perspectives and conclusions that are often obscured in traditional histories, which tend to focus on abstract concepts and ideas. Think, for example, of the role that natural resources such as abundant land, timber, and animals such as the beaver had on the growth of business in the nation. Consider how new forms of artificial lighting—whale oil, natural gas, and ultimately electricity—changed patterns of work and production. Ponder how increased desire in the United Sates for variety and customization in consumer goods has fueled development of new businesses, both in this country and around the world. Consider how a complex contemporary object, like a handheld digital telephone, contains within itself the results of centuries of innovation and product development. Its materials, design, and functions encapsulate the outcomes of thousands of decisions and

the creative work of scores of individuals. Artifacts like this are time capsules of their own historic development.

Over two centuries ago, George Washington sat at Mount Vernon, wondering what the nation he helped create would become, and whether its commerce and trade would rank among those of other leading countries. Certainly he could never have foreseen how continual innovation would ultimately undermine the merchant era in which he lived, and how it would be followed by corporate, consumer, and global eras. But perhaps as surprising to him as the changes that were to come would be the endurance of the fundamental principles and ideas that he helped establish to foster the growth of the nation: a national commitment to capitalism and democracy; and a social commitment to individual opportunity and the common good. These shared values—and the dynamic tension among them—have animated American enterprise ever since.

DAVID K. ALLISON
Associate Director for Curatorial Affairs,
National Museum of American History

View on the Erie Canal, by John William Hull, 1830, watercolor on paper. The opening of the Erie Canal in 1825 gave access to the Western interior, aiding the flow of imported, artisan, and factory-made goods headed west from the Port of New York and to produce

The Merchant Era
1770–1850s

THE UNITED STATES was born in the Merchant Era (1770–1850s). In 1790, the nation's population numbered 3.9 million people not including Indians: four-fifths of them were of European background; the other fifth, African. The country included 864 thousand square miles of land stretching along the East Coast of North America: from the Atlantic Ocean in the East, to the Appalachian Mountains in the West; from Massachusetts in the North, to Georgia in the South. It was sparsely populated and rich with resources, including farmland, timber, coal, iron ore, and minerals. In large part, the War for Independence had been a fight over who would benefit from this enormous potential.

Americans, including the immigrants now flooding in, were anxious to seize the opportunity. A majority of them made their livings on farms in rural areas. Some were tradesmen or merchants in towns or the few cities. Around 700,000 enslaved people worked mostly on Southern farms and plantations. Overall, Americans had a higher standard of living than inhabitants of Great Britain, but wealth was unequally distributed. Historians estimate that the top 10 percent of the population controlled between 40 percent and 60 percent of the wealth. There was little gold or silver money for buying goods. Instead, Americans bartered, or used skins, tobacco, shells, and paper bills. Although widely dispersed over vast areas, Americans did not live isolated, simple lives.

He That Tilleth His Land Shall Be Satisfied, **artist unknown, circa 1850, oil on wood.** Agriculture was the economic mainstay of the nation at this time. Farmers harvested their fields of grains and tobacco to exchange in the marketplace for needed goods.

The nation was experiencing what—in retrospect—historians call the "market revolution." It was a fundamental change in how goods and services were produced and consumed and how business was transacted. People bought or traded for goods in a physical and social marketplace, but they also acquired objects that were brought to their rural world from across the globe. These came to them from peddlers who met them at their door or from merchants who stocked these items in their stores. Merchants sold handkerchiefs from India, ceramics from England, and tea from China. Fur traders in the Western territories purchased beads made in Amsterdam or Venice. One of the most lucrative businesses was the most horrifying—slave trading. It ranked among the largest and most capital-intensive enterprises in America until slavery was abolished.

New kinds of transportation provided more opportunities to buy goods from distant producers. Canals, turnpikes, steamboats, and even simple carts carried an array of merchandise to local merchants. Markets were no longer tied to a physical place, nor did exchanges occur only when one shook hands or looked another in the eye. Unbound by time and space, the market became an activity dependent on the presence of goods, often taking place over long distances with less personal acts of buying and selling. Financial instruments such as bills of exchange or private banknotes provided an additional layer of opaqueness and detachment. The securities of a household economy were giving way to the risks of a rising commercial capitalism.

With the introduction of new industrial technologies and systems of production by Samuel Slater and other entrepreneurs, lifelong work practices dissolved. Many workers who had previously labored at home or in small shops, with a variety of simple machines and tools, now toiled in larger, centralized shops with more specialized machinery, new industrial rules, and increased managerial control. Some craftsmen and artisans built small enterprises into larger and long-lived companies. With all these ventures, while objects such as textiles and shoes changed in form, Americans' lives as producers and consumers were being transformed.

Embargo Act miniature teakettle, metal and copper, 1807–1809. This tiny teakettle, likely made to demonstrate an artisan's skills, expressed sentiments responding to Thomas Jefferson's 1807 Embargo Act. The engraving of "mind your business" admonished the trade person to focus and work with diligence.

Handkerchief, 18th century. Handkerchiefs from India often appeared on merchants' ledger accounts. A fashionable and practical accessory for men and women, a handkerchief could be tucked in a pocket, worn about the neck, or carried in the hand. Ramsay lists an imported "Sastracundy" handkerchief, the name identifying its Southern Indian origin.

These changes are well illustrated in the lives of two merchants from the era: William Ramsay and Polly Salmon, who found new opportunities to do business, make money, take risks, and purchase consumer goods as the market revolution changed the way Americans engaged in trade. As the transformation spread, opportunities arose from westward expansion into land acquired from Native Americans, from expanding trade routes overseas to China, and from new innovations, such as artificial lighting, which forever changed the structure of workdays.

BUYING AND SELLING: TWO MERCHANTS

Born in Scotland, William Ramsay (1716–1785) settled in Virginia in the 1740s. He lived most of his life as a colonial merchant under British rule. Early in his mercantile career, Ramsay realized that fortunes could be made at the edges of the frontier near the Potomac River. In 1748 he and several others petitioned the Virginia General Assembly to found a new town at Hunting Creek Warehouse. The Assembly approved, noting "it would be

Samuel Slater's indenture contract, 1783. His indenture in an English cotton mill completed, and intrigued by the possibilities of work in America, Samuel Slater left England in 1787 disguised as a farmer. Soon after his arrival on these shores, he built the first effective American spinning mill.

commodious for trade and navigation and greatly tend to the ease and the advantage of the frontier inhabitants" (Ramsay 1999, 3). They also required that Ramsay and his fellow petitioners allot a portion of the land for a marketplace and a public landing. The new town was named "Alexandria."

The detailed records in Ramsay's nearly nine-hundred-page ledger book of 1753 to 1756 confirmed that the enterprise greatly benefited his mercantile activity. The pages document how the country's farmers, tradespeople, and gentry became consumers, debtors, and creditors. Trade occurred across gender, occupation, and race. Enslaved peoples bought cloth; women purchased shoes; men acquired ribbons, buckles, and farmers, shoes. Even George Washington was a customer—buying a dozen silver buttons and a pair of men's glazed lamb gloves.

Later nineteenth-century accounts verified that Ramsay's familiarity with the "markets of the world and the laws of the trade" made him popular with farmers on the frontier, who brought him their produce and purchased his supplies. Ramsay's position in the marketplace was further strengthened by the public office he held. He became adjuster

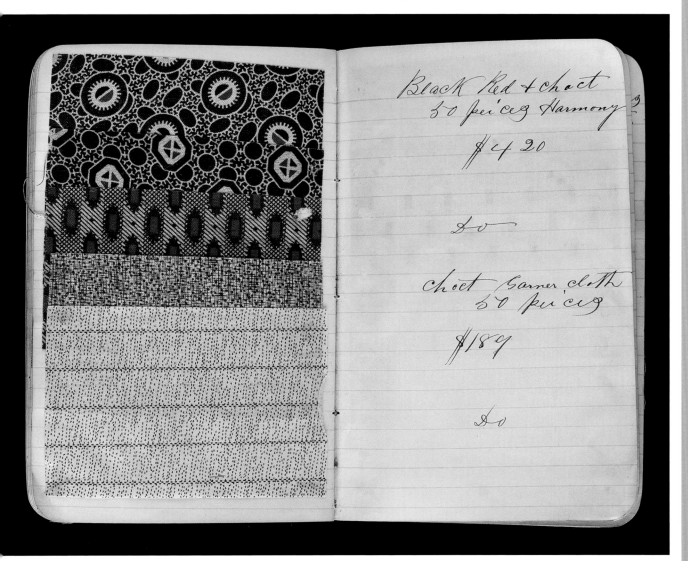

A dyer's sample and cost record book, 1877. With new technologies in textile production, dyers endeavored to meet manufacturers' demands for exact shades by matching color lots. John Oldfield kept this sample book as the boss dyer at the Garner Print Works in Wappingers Falls, New York.

WATCHMAKER SIGN

Before the advent of advertising agencies, merchants and shopkeepers did their own advertising. Artisans and tradespeople used images to inform a public that was mostly illiterate of their wares. Signs with pictures and symbols identified the sorts of goods that might be found in any given shop. Typically such signs, often made by itinerant artists, needed to be large enough to be seen from a distance, bold in shape, eye catching, and easy to remember. This watchmaker sign did the job, broadly identifying the Barthelmes shop. Inside the store, it was likely, though, that in addition to pocket watches a much broader assemblage of goods such as jewelry, clocks, and optical wares would be found.

Robert Donald, **18th century.** This successful Scottish merchant lived in Alexandria, Virginia, and knew William Ramsay. There are no known images of Ramsay. Rather than remain in the colonies, Donald chose to return to his native country once he made his fortune here.

Barthelmes watchmaker sign, wood, mid-19th century.

of the weights and seals and arbiter of the dry, wet, and weight measures. With these tools, Ramsay determined the accurate quantity of farmers' produce. The stability of the marketplace depended on the fairness of trade and trust in those who ensured it. Regulation protected the consumer, the producer, and the merchant.

The market revolution began with transactions at the merchant's desk. Like many surviving desks and bookcases of the eighteenth century, Ramsay's desk was equipped as a business center, at which most of the merchant's financial activities took place. His account ledgers fit the niches in the upper portion of the bookcase, where they were easily accessible. Here he wrote and received letters from George Washington, Patrick Henry, and Benjamin Rush concerning provisions for the Revolutionary War, navigation of the Potomac River, and shipping of goods from England. Ramsay's papers and his

William Ramsay's desk, bookcase, ledger, weights, and measure, 18th century. Functional as well as decorative, William Ramsay's imposing desk was indicative of his success as a merchant. Likely the doors stood open to display the elaborate interior. Ramsay's marketplace weights and measure are at the ready.

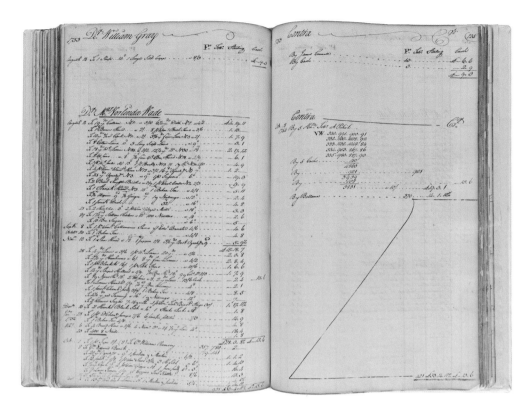

William Ramsay's ledger, 1753–1756. William Ramsay kept this parchment-covered ledger on his desk to record the activities of his store in Alexandria, Virginia. On the left hand page buyers' names, dates, purchases, and prices appeared; the "contra" side noted the date of remuneration, the amount, and method of payment.

Chocolate pot and whisk, 18th century. Many American merchants sold chocolate. Consumers enjoyed hot chocolate, either in their homes or in a public chocolate house. Most often it was served in a special pot, like this one, with a lid that had an opening in it for the whisk.

above, left: **Earthenware mug, 18th century.** William Ramsay owned the *Fairfax*, a vessel he unsuccessfully tried to sell in 1751. The ship had its own ledger account that listed this mug and other provisions needed to maintain the vessel at sea.

above, right: **Man's knee buckle, 18th century.** Both men and women purchased fancy buckles to adorn shoes and clothing. This man's buckle has twenty-four paste (cut glass) stones set in steel. William Ramsay's ledger shows one "pair knee buckles" sold to a John Douglass of Virginia.

will, now in the collection at the National Museum of American History Archives, likely were written at this desk. In his 1785 will, Ramsay bequeathed all his furniture to his wife, with one exception itemized first: "…I give and desire my mahogany Desk and Book Case unto my son William…" (William Ramsay's Will, Smithsonian Institution).

Boston milliner and merchant Polly Salmon made the first entry in her memorandum and expense book in January 1785, the year William Ramsay died. In this era, women were active as merchants and shopkeepers, particularly in coastal urban centers. There, imported goods were readily accessible and women were highly knowledgeable about consumer preferences. Salmon may have financed her business with the sale of property. Newspaper accounts in December of 1784 record that Salmon's house and land—No 22 Cornhill—sold at public auction. According to the advertisement, the land was valued because it was near the town pump, had been improved by Salmon, and—as noted in Boston's *Continental Journal and Weekly Advertiser* of December 16, 1784—was well known for many years as the best stand for a shopkeeper in the town.

It is unclear why she sold a property that was advantageously located for shopkeepers. However the entries in her memorandum book beginning three weeks after the auction indicate that she was still actively involved in business. She purchased shoes from merchant John Upton, and three days later paid Francis Amory fifteen pounds for unknown merchandise. Perhaps Salmon assumed a new role as middle person, purchasing goods from large mercantile firms and reselling to small shopkeepers. Like all merchants, she took risks in purchasing articles that might be fashionable one day, and out of style soon after. Or she wrongly anticipated the pulse of the market and was left with quantities of unwanted stock. Her little memorandum book recorded such activities and probably accompanied her as she made her rounds from merchant to merchant, with each entry in the hand of the seller. These notations were likely cross-listed to another account book.

PAUL REVERE
1734–1818

His name is more often associated today with his hasty midnight ride to Lexington and Concord, Massachusetts, during the Revolutionary War, but that was not how Paul Revere's contemporaries viewed him. A master silversmith and owner of a silver- and goldsmithing shop, Revere was admired by Bostonians who valued the quality of his craftsmanship. He had learned the trade from his father, Apollos Rivoire, a French Huguenot immigrant goldsmith. However, like most artisans of the period, Revere could not rely solely on one trade to earn a living. To supplement his income he also labored as a copper plate engraver and as a dentist.

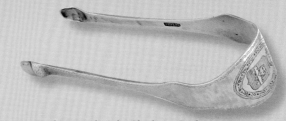

above, top: **Paul Revere**, by John Singleton Copley, 1768.
above, bottom: **Paul Revere's silver sugar tongs, 1792.**

STEPHEN BURROUGHS
1765–1840

In the early nineteenth century, Stephen Burroughs' reputation as a counterfeiter extended beyond the mere replication of paper money—which he was very good at—into the larger realm of the imposter and impersonator. His best-selling autobiographical memoir laid out his life of aliases as a school teacher, library founder, seaman, and minister as well as his Houdini-like escapes from prisons. The counterfeiting of his character indicated the changing nature of Americans' conception of self—as a self-made person in a rapidly changing commercial world. To counteract Burroughs' notoriety, moral reformers used his waywardness as an example of a "woefully ill-spent" life.

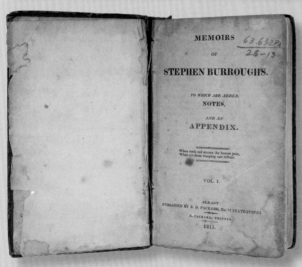

above, top: **Counterfeit bill, 1807.**

above, bottom: ***The Notorious Memoirs of Stephen Burroughs,*** **vol. 1, published by B.D. Packard, 1811.**

Salmon procured goods from several large mercantile firms that advertised extensively in the Boston and regional newspapers. On November 23, 1785, she paid John and Daniel Jenks in full for four pounds four shillings worth of ribbons— a goodly number of ribbons! Salmon acquired the most up-to-date styles, for the Jenks gave notice in newspapers like the *Salem Gazette* in November of 1784 and May of 1785 that they imported from London a "Fresh Supply of GOODS," including ribbons and all sorts of other fancy items.

Men and women in this era, enslaved as well as free, used ribbons for many purposes. Ribbons made fashion statements. They were used as trim to enliven frequently worn clothing, knotted at the neck to suspend jewelry, tied around one's hair, or worn close to the body to draw in a stomacher. "Love ribbons" were given to favorites, while black mourning ribbons were given to family members in grief.

Polly Salmon's memorandum and expense book, from 1785 to 1788. Ribbon, late 18th century. The diminutive size of Salmon's record book alludes to its portability. Entries are in different hands, suggesting that merchants signed their own names to confirm payment. Within the book, Salmon accounts for several purchases of ribbon.

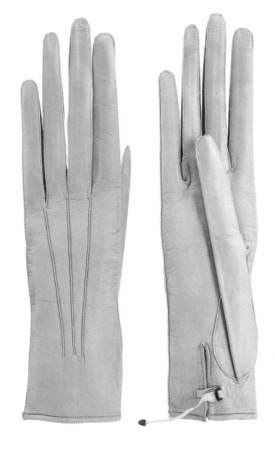

above, top: **Gloves, early 19th century.** In 1786 Polly Salmon noted in her memorandum book the purchase of three dozen pairs of women's gloves from merchant John Blanchard. In the Boston newspapers a John W. Blanchard advertised wares in his store at #10, Long Wharf.

above, bottom: **Shoes, 18th century.** Polly Salmon and William Ramsay imported shoes for their customers. In August 1753, Vorlinda Wade purchased eight pairs of "best shoes," from Ramsay. Wade was a widow and the sole owner of a 193-acre tobacco farm that adjoined George Washington's Alexandria, Virginia, property.

Salmon's business was wide-ranging. She traded with Perez Morton, a wealthy merchant, lawyer, and later Massachusetts Attorney General, as well as Boston female merchant Susannah Renken. In 1786, she bought ten pairs of cloth shoes from John Upton, three dozen ladies gloves from John Blanchard in 1786, and "one pound sixteen shillings worth" of handkerchiefs from the Brimmers.

Of all the merchants in the memorandum book, Salmon purchased the greatest amount from Andrew Brimmer. After twenty-three visits to his establishment, she knew him well enough to accept his proposal of marriage in February 1788. The Boston paper noted, "Married—At Boston, Mr. Andrew Brimmer, merchant, to Miss Polly Salmon" (*Salem Mercury,* February 26, 1788, 3). Salmon's last entry in the memorandum book was February 16, 1788—four days before her marriage. Though the memorandum book ends, it is likely that the capable Polly Salmon Brimmer continued to work beside her husband.

BREAKING A MONOPOLY: THE RED RIVER CART

Fur trading was an extensive international business during the Merchant Era. To a large extent, it was controlled by European organizations, notably the English Hudson's Bay Company. But it depended on furs from local suppliers in America, many of them American Indians. Prominent among these were the Métis, a mixed European-American and Indian people. They used innovative, hand-crafted vehicles to convey furs hundreds of miles south to St. Paul in the Minnesota Territory.

Each cart was constructed differently, yet all shared similar characteristics. Made entirely of wood, including the wheels, the members were lashed together with raw hide that tightened as it weathered. Bison hides covered the cart, protecting the furs. The skins were also used when the cart was flipped over to function as a raft when fording a river. In those situations, the wheels were placed flat side by side, the wooden frame placed on top, and the

Trapping Beaver, by Alfred Jacob Miller, 1858–1860, watercolor on paper. In 1837, the artist Alfred Jacob Miller accompanied fur trappers on their hunt for beaver in the far West. Miller's field notes and sketches described locating the beavers and setting the five-pound traps. Miller's sketches resulted in finished watercolors.

Beaver pelt, 19th century. Thick beaver fur was the most desirable. Three grades of pelts defined the trade: *castor gras*, pelts worn by trappers whose perspiration and body oils softened the coarse outer hair; *castor sec*, stiff unworn scraped pelts, and *bandeau*, poorly scraped pelts.

Beaver hat, 19th century. Beaver hats were fashionable for over three hundred years. The quality of a beaver hat indicated position and wealth. Pelts traveled by cart from the Dakota Territory to St. Paul, then on to the East Coast and Europe, where they were fashioned into hats.

hides enfolded the cart. Such flexibility was necessary. Almost always, these carts took paths following rivers that led to distant markets. They were named Red River carts for the river that flowed through the upper Midwest and into Canada, where many of them were used.

Imagine a vast train of Red River carts, often five hundred or more, noisily squeaking their way across the plains. Such a far-reaching trade occurred as the result of trade inequities at the cart's source, in the Dakota Territory's Red River District. To maximize their profits and limit competition, the Hudson's Bay Company tried to control the supply of furs. In 1849 a company representative arrested Métis traders for transacting business with noncompany sources. Frustrated with the restrictions and monopolistic practices of the company, the Métis rose up and demanded the release of their associates. The company's attempt to intimidate them backfired, when the court dismissed their case. The Métis declared, "commerce est libre!" when they heard the decision. They won the right to trade freely with whomever gave them reasonable prices for their furs and fair rates for the goods they wished to buy. By 1856, nearly half the commerce from the Red River District had shifted to merchants in St. Paul.

Métis and Indian women were in the forefront of their tribes' and families' fur-trading business. Women sometimes trapped, tanned hides, or hunted small game and often negotiated the terms for the sale of goods. This role as liaison between the exchange of furs and access to trade goods enhanced their authority within the community. In 1858, Jane Grey Swisshelm, editor of the *St. Cloud Visiter* newspaper and a forceful advocate of women's rights, observed the Métis in St. Cloud, Minnesota. The town served as a way station for them on their route to St. Paul; it was here that many camped and then crossed the Mississippi River on their way south. Because they spent time in the town, Swisshelm had an opportunity to observe their practices. She remarked: "The carts of the women are painted; and have a cover with other appearances of greater

JOHN JACOB ASTOR
1763–1848

John Jacob Astor likely never set a beaver trap, though he made a lucrative living selling their fur. Learning the ways of business from his father, a butcher in Germany, and then from one of his brothers living in England, Astor arrived in the United States in 1784, where he applied his acumen to a fur-trading business in New York. Taking calculated risks, he expanded his small trade into the continental American Fur Company, with trading posts from the Great Lakes to the Northwest. While trading furs in China, Astor recognized a more profitable market of exchange in Turkish opium. Astor sold his assets in the fur trade and invested considerable funds in Manhattan real estate. At his death he was the wealthiest man in America.

John Jacob Astor, by John Wesley Jarvis, circa 1825.

A Half-cast and his two Wives, by Peter Rindisbacher, about 1825. Métis women moved between two worlds—the Indian and the European. As capable traders, the women knew the value of the furs they sold and the merchants' goods they wished to buy.

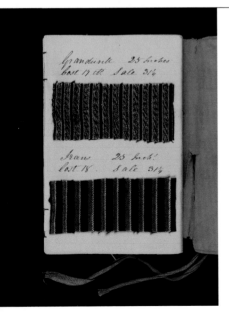

above: **Green River knife, 1840.** A great admirer of English Sheffield cutlery, John Russell was determined to provide Americans with a knife of equal quality, and he did. Named after his factory on the Green River in Massachusetts, this all-purpose knife became the standard, and revered, frontier tool.

far left: **Indian trade goods, fabric sample book, 1834.** This book listed the prices of cottons of Eastern manufacturers compared to the selling cost to the Indian. It provides insight into the merchants' costs and markups, and offers a rare example of cloths and the terms assigned to them.

left: **Glass bead necklace, Plains Cree Nation.** Glass beads made in Venice or Amsterdam were widely purchased by the Métis and Indians and came to replace traditional decorative materials, such as porcupine quills, on garments and moccasins. Their reflectiveness and solidity had symbolic meaning to the Indian.

attention to comfort than is displayed in the carts appropriated to the men" (Gilman et al 1979, 15).

Once in St. Paul, the women bargained with the local merchants for the goods, many of them from distant producers. Purchases included knives, tin containers, muskets, beads, kettles, and cloth. After a respite of three weeks at the end of summer, the carts loaded with goods began the long trek back to the Red River District.

The Smithsonian's Red River cart has a noteworthy history. It was donated in 1882 by Charles Cavileer, the first customs officer and postmaster of Pembina, Dakota Territory. According to a commentary in the *Grand Forks Daily Herald* in 1892, Cavileer traveled through the territory in the cart nearly forty years earlier. In the 1880s, numbers of Red River carts, or portions of them, were strewn across the Dakota landscape—vestiges of an earlier trade. Cavileer recognized that they represented a history of commerce that deserved preservation. He disassembled the cart and sent it to the Smithsonian with the hope it would record a culture and an activity that was fast disappearing. It became the very first vehicle accepted into the young institution's National Collection of transportation artifacts, and was displayed soon after. A decade later, the cart traveled to the 1893 Chicago World's Fair, where it represented the state's history in the North Dakota building. Subsequently, the vehicle sat quietly in storage, until its recent conservation for the American Enterprise exhibition.

Red River cart, circa 1840. Though portions of the cart are missing or badly decayed, it remains a rare survival of a once ubiquitous vehicle of the northern plains. Generally the cart carried only furs and trade goods. Its large wood wheels and design did not provide a comfortable ride.

THE BUSINESS OF SLAVERY

In the Merchant Era, slavery created enormous profits for businessmen in both the North and South. It was an important part of the market revolution. With the closing of the African slave trade in 1808, domestic commerce in slaves escalated. At that point, most slaves lived in the Chesapeake region, but the need for their work in tobacco fields had greatly lessened with the soil's depletion. Consequently, slave owners looked for a new market for their surplus hands. They found it in the lower South, where the rise of cotton created a growing need for labor.

Enslaved adults and their children were legally designated as property. They could be acquired and conveyed to any place where slavery was permissible, for any purpose. An asset in slaves was seen as a way for upper and lower Southern owners to advance their fortunes. Slave-trading firms encouraged buying early and often: "…Call and make your purchases to gather your crop (slaves)—and then call quick again and buy to make another crop. By those means if you will keep up your purchases for ten years there is no telling how much you may be worth. This is the true Road to wealth…" (Deyle 2005, 94). Northern merchants supported slave traders by providing insurance contracts and other negotiable paper to facilitate

Slave Auction, Christiansburg, Virginia, **from Lewis Miller's sketchbook of landscapes in the state of Virginia, circa 1853, watercolor.** Pennsylvanian Lewis Miller viewed this public sale of slaves when he visited his brother in Christiansburg, Virginia. The slaves' names may substantiate the accuracy of the scene.

FORREST & MAPLES slave sale advertisement, 1855. Nathan Bedford Forrest ran a slave-trading business in Memphis, Tennessee, accruing a fortune. As remembered by slaves, Forrest's establishment had two small "Negro" houses surrounded by a stockade where slaves paraded on auction days.

their trade. This was in addition to cotton mills and other businesses that made money from cotton trading. With cotton as one of the principle sectors of the U.S. global economy in the first half of the nineteenth century, and enslaved labor at its fulcrum, the whole nation bore culpability for the institution of slavery.

ELIZABETH KECKLEY
1818–1907

Elizabeth Keckley was born a slave in Virginia. A seamstress and dressmaker and skilled in sewing, even at a young age, she would use those talents to earn money that thirty-four years later bought her freedom. She, like many other slaves, resisted enslavement. As a free woman Keckley moved to Washington, D.C., where she found well-placed clients interested in her sewing abilities. Mary Todd Lincoln employed her as a dressmaker, and later accepted her as a trusted companion. With savings from her work as a dressmaker, Keckley founded the Contraband Relief Association to assist newly freed slaves. She made this christening gown for her goddaughter.

Christening gown, 1866.

NEWSPAPER ADVERTISING

Almost from their beginning in America, newspapers carried advertisements, and often the commercial announcements made up a large section of the paper. In Southern newspapers, it was not uncommon to find an advertisement for coffee, Jamaica rum, horses, and slaves all in the same block of text. As a business, the slave trade was dependent on sales, with notices by traders offering to purchase slaves, others posting slave auction dates, and still others identifying available slaves for sale. Many advertisements were for runaway slaves, with extensive text describing the details of their dress, their skills, and their ages.

The Columbian Mirror and Alexandria Gazette, March 6, 1793.

Alexander Whilldin exemplified the Northern businessmen who benefited from slavery. When he retired in 1871 as president of the American Life Insurance and Trust Company of Philadelphia, after sixteen years with the firm, Whilldin was lauded as a citizen of the most eminent reputation. His standing ensued not only from his work in the insurance business, but also from his ownership of a large cotton and wool commission house. Although Whilldin was not directly involved in the slave trade, his livelihood depended on it, as he profited from both selling cotton and insuring slaves. Consider the story of a slave named William.

In 1858 William, worth $900, lived in Charles City, Virginia, as the property of Monroe Franklin Vaiden. Vaiden was thirty-seven years old and a merchant whose principal assets lay in his store, in William, and possibly in William's wife Rosanna, valued at $600. To protect himself from financial loss if his slaves died, Vaiden insured them with the American Life Insurance and Trust Company. Dr. Charles R. Bricken of Richmond, one of the insurance company's three Virginia agents, visited Vaiden in April 1858 to sell him a policy. Unlike other firms, the Philadelphia company refused to insure against slaves escaping, being kidnapped, or dying from "neglect, abuse or mistreatment by his/her owner" (Vaiden 1858). Records indicate that William's policy cost $16.20 per year. This and other transactions like it were good business for both Whilldin and Bricken.

Slavery touched many areas of the economy. To increase efficiency, traders used printed and stamped standardized forms. The forms were often handsomely designed and embellished, belying their intent. No space was provided for individual human details or characteristics. Take, for example, an 1850 receipt issued by Jones & Matthews in Richmond, Virginia, that noted the sale of slave Nate to J.C. Sproull for $850. All human context is absent. Lacking detailed records, we can only guess Nate's fate. Since Sproull served as a contractor for the South Carolina Canal and Railroad Company, Nate might have done several years of hard labor in roadbed construction. In April 1852, Sproull advertised the sale of their 130 "negros," 85 mules, and 90 carts and harnesses in Aiken, South Carolina. Was Nate one of them? The notice stated: "These Negroes are beyond doubt the likeliest gang, for their number, ever offered in any market, consisting almost entirely of young fellows from the age of 21 to 30 years, and a few boys from 12 to 16 years of age..." (*Times Picayune*, 1852, C3).

Tax receipts show another dimension of the slave trade. Property taxes were important sources of revenue for most Southern states, and they included taxation of slave valuations. By 1860, slaves in the South were worth more than three billion dollars. It is unclear whether Andrew Souillan, the person noted on the tax receipt shown here, owned slaves or contributed to this tax base. Nonetheless, the receipt is illuminating. Souillan's tax levy of $150 was equal to the valuation of one sixteen-year-old slave.

No part of the slave business was more heart-wrenching than the interstate trade. Historians calculate that as many as half of all slave families of the upper South were torn asunder by it. Surviving records sometimes record the ages and

Andrew Souillan's tax receipt, November 1835. Tax collected on slave sales, on property that included the value of slaves, and on estates served as significant sources of revenue for state and local governments.

full names of those involved. Between December 1832 and early January 1833, for example, slave trader John Armfield took eighty-three enslaved individuals from upper South plantations, and

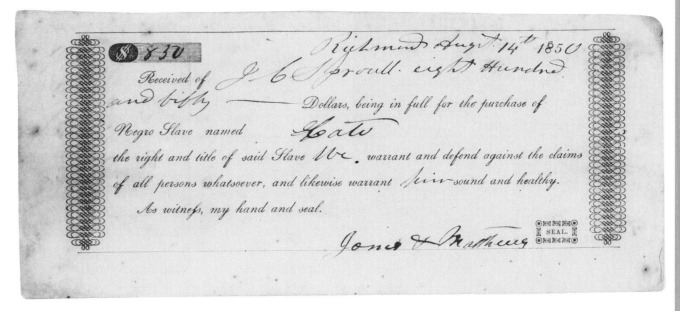

Receipt of slave sale, issued to J.C. Sproull, by Jones and Matthews, Richmond, Virginia, 1850. The business of slavery included the production of standardized printed forms for the slave trader's use. Typically such forms just proved that a transaction took place; this one concerned the sale of a human being.

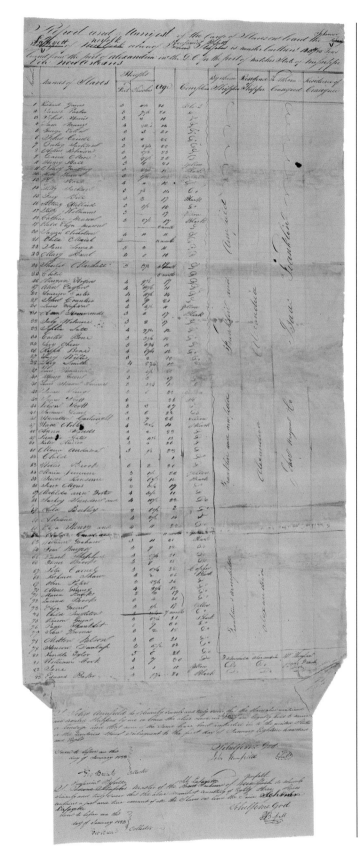

placed them on the schooner *Lafayette* bound for New Orleans and Natchez. They included Moses Pinney, age twelve (#70), and Matilda Anne Foster, age eleven (#57). How did their mothers react? No record survives, but in a similar instance, a mother protested so vociferously that she was lashed severely on her shoulders. Slaver Armfield was once asked how slaves responded when separated from their families. He brazenly stated, "Sometimes they don't mind it a great deal; at other times they take on right smart, for a long time" (Tadman 1989, 167).

Armfield and his partner Isaac Franklin operated the largest slave-trading operation in the South. They managed their business out of Alexandria, Washington, D.C., New Orleans, and Natchez in an innovative and efficient manner. Armfield ran the upper South part of the business. After assembling individuals for sale, he housed them in his Duke Street complex. It included a large yard, a hospital, a tailor shop where slaves were outfitted for the slave block, and a kitchen. In addition, Armfield ran their firm's coastal shipping operation, transporting slaves to Mississippi on their own vessels. This vertically integrated business strategy eliminated middlemen and increased profits.

Franklin managed the selling end of the operation. Because Louisiana law restricted the import of slaves for sale after 1832, Franklin lived over the border in Natchez, Mississippi. Aided by relatives and a network of purchasing agents, Franklin bragged in 1833: "[I] sold more Negros than all the Traders together" (Deyle 2005, 104). To promote sales he enhanced the appearance of the slaves with new clothes, dyed hair, and other embellishments. Potential buyers prodded and examined the enslaved, carefully trying to find the best deals. Slave traders paid close attention to profits and losses. They often used double-entry bookkeeping, and preferred cash for sales. Only when pushed did they sell on credit or take promissory notes.

Slave manifest of schooner *Lafayette*, 1833. This ship manifest recorded the January day when eighty-three slaves of the upper South embarked on a journey to an unknown location and undetermined fate in the Deep South. It was likely they would never see their families again.

Though widely known to be slave traders, Franklin and Armfield led respectable lives in the eyes of most of their contemporaries. One said of Armfield, "[he] bears a good character, and is considered a charitable man" (Deyle 2005, 138). Amassing huge fortunes, the two endowed a southern college and an institute. While thousands of other people involved in the business of slavery never achieved the wealth of these men, they all benefitted from this reprehensible element of the market revolution.

REDISTRIBUTING LAND

During the Merchant Era, the territory of the United States expanded across the continent through a variety of treaties, purchases, and negotiations. The Treaty of Paris in 1783, which formally ended the American Revolution, established the original boundaries of the nation. The Purchase of Louisiana from France in 1803 doubled United States territory. Florida was bought from Spain in 1821; Texas annexed in 1845. The Mexican War added the Southwest and California. The Oregon Territory was annexed through a treaty with the British in 1846, and the Gadsden Purchase from Mexico completed continental expansion in 1853.

These well-known land acquisitions from other nations, however, were far from the full story, for living within the new lands were both Native Americans and others, such as the French and Spanish settlers, who had land claims that would have to be settled. After the U.S. government made the treaties at the national level, additional negotiations took place with these inhabitants. Two examples are treaties with the Kansa and Menominee tribes in the 1820s and 1830s.

Chief Sho-me-kos-see of the Kansa nation lived in the central Midwest, and Chief Makakapaness and his Menominee nation in present-day Wisconsin. Both nations had supported the British during the War of 1812, but both signed their first treaties of friendship with the United States soon after the war ended. Their chiefs received peace medals as

JAMES DeWOLF
1764–1837

Learning the shipping trade as a successful privateer during the War of 1812, James DeWolf applied these skills to become a notorious slave trader, bringing slaves from Africa to the auction blocks of America. He defied the 1787 Rhode Island restriction, and the later 1808 United States law forbidding the importation of slaves, by evading customs inspections and using Cuba as a slave depot. His position in the Rhode Island General Assembly from 1802 to 1821, and his election as a United States Senator in 1821, solidified his standing. The commerce in slaves, along with the development of his Rhode Island cotton mill, brought him great wealth and political prominence in his lifetime, though little esteem today.

James DeWolf, artist unknown, 1825–1830.

Fort Pierre, Mouth of Teton River, 1200 Miles above Saint Louis, by George Catlin, 1832, oil on canvas. George Catlin painted this view before the arrival of settlers. The painting shows the Sioux encampment with its six hundred skin lodges. Here Indians traded furs at the American Fur Company post.

tokens of the new relationship. By the mid-1820s and early 1830s, white settlers were moving west into the land the United States had acquired through the Louisiana Purchase, and the tribes were forced to adjudicate their property claims.

William Clark, co-leader of the Lewis and Clark expedition, was named Superintendent of Indian Affairs in the Louisiana territory in 1807, a post he would hold until 1838. Negotiating with the Kansa fell under his jurisdiction. In a treaty in 1825, he convinced the tribe to accept $4,000 in livestock and goods, $3,500 annuity for twenty years, and a reservation along the Kansas River, in exchange for any claims to what is currently half of the state of Kansas. Eleven years later the Menominee gave up their claims to lands in northeastern Wisconsin in exchange for $20,000 in cash, $3,000 in food, several thousand pounds of tobacco, $500 for agricultural equipment for twenty years, and settlement of $99,000 in debts.

Although the Menominee continued to live in north central Wisconsin, those residual lands soon came under pressure as well. Initially agreeing to move to Minnesota, they successfully petitioned President Fillmore in 1852 and won the right to remain in Wisconsin, on a 250,000-acre reservation.

The Kansa, too, reached out to an American president. In 1863, one year after the disastrous Dakota War in southwest Minnesota, they addressed a letter to President Lincoln. They laid out their grievances: thieving Indian agents, delinquent and missed federal payments, and poor land. They begged: "My Great Father! White men tell us that you are going to drive us off to another place. We don't want to go. We want our children to have this place when we are dead" (Petition of Is-tata Sin and others to Abraham Lincoln, 1863). Forced off their land along the Kansa River, the tribe was ultimately relocated to Oklahoma in 1873.

As white settlers sought ever more land in the South as well as the West, Indian removal expanded and became more systematic. In 1830, the federal government passed the Indian Removal Act, which authorized the relocation of five tribes from the South to designated reservations in the West: the Chickasaw, Choctaw, Muscogee-Creek, Seminole, and Cherokee. Historians would later call this sad episode the "Trail of Tears."

In 1839, topographer J. Goldsborough Bruff summarized years of treaties, negotiations, and legislation into a comprehensive Indian Land Cession Map. This remarkable document detailed the fate of land claims of more than twenty-four Indian tribes, including the five above, the Kansa, Menominee, and others. The legends show the acreage they held, as well as the dates and locations where their treaties were signed. Likely this map guided the U.S. Congress as it planned future Indian removals and new settlements. However, its seeming clarity masked a violent clash of cultures and ideologies of the parties who had presumably agreed to the treaties. They had radically different

above, left: **Sho-me-kos-see, The Wolf, A Chief, by George Catlin, 1832.** Sho-me-kos-see, a chief of the Kansa nation from the central Midwest, wore his peace medal with other ornaments when George Catlin painted him in 1832. According to Methodist missionary Reverend William Johnson, Sho-me-kos-see was the only Kansa to convert to Christianity.

above, right: **James K. Polk peace medal, 1845.** Said to have been owned by Menominee Chief Makakapaness, this medal affirmed the friendship and sovereignty of the United States with his nation. The reverse side shows hands clasped in goodwill and a tomahawk crossed with a peace pipe.

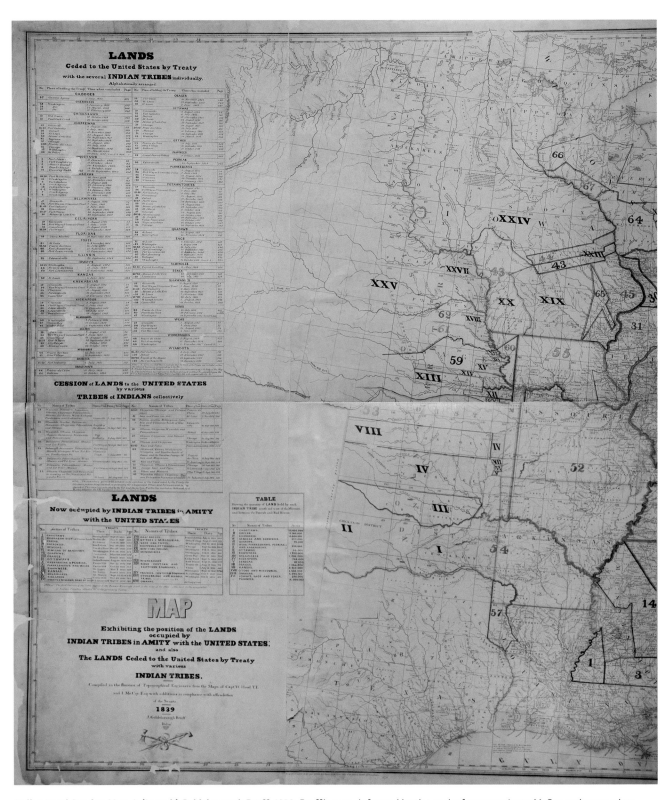

Indian Land Cession Map, J. (Joseph) Goldsborough Bruff, 1839. Bruff's map, informed by the work of surveyor Isaac McCoy and mapmaker Captain Washington Hood, apprised the United States government on lands and acreage held by Indian nations. Bruff's map assisted Congress as it funded the 1830 Indian Removal Act.

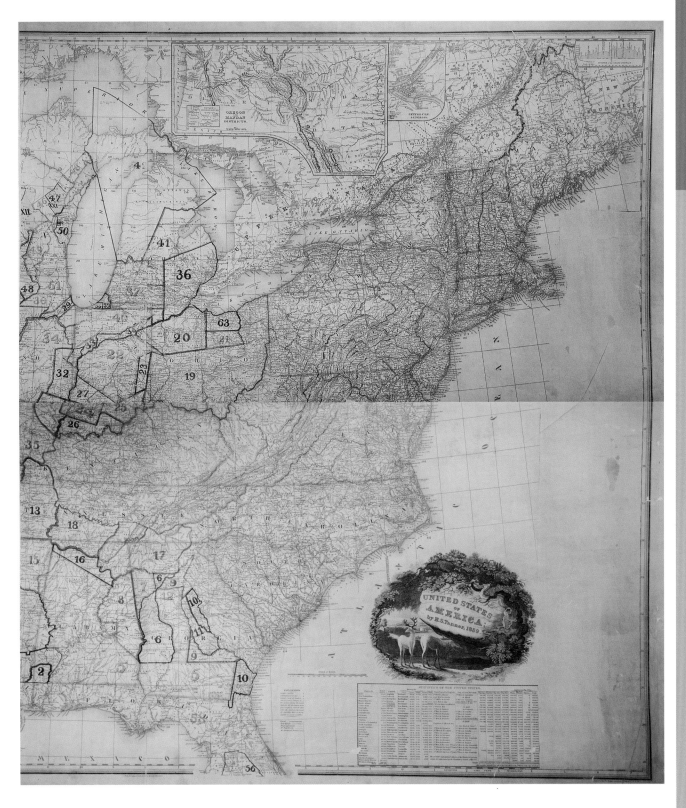

understandings of the right relationship between people and the land that sustained them, and what the concept of "ownership" meant. These fundamental disagreements would reverberate through American history far into the future.

With most Indians removed, land was ripe for speculation. Following a continuing pattern, business people saw opportunities to develop the region and encouraged settlers to migrate west. While many settlers and settlements succeeded, others failed. In 1851 John Nininger established the township named for himself on the eastern bank of the Mississippi River, thirty-five miles below St. Paul. A local businessman and speculator, Nininger made a concerted effort to appeal to eastern investors and immigrants. His broadsides touted the advantages of the land and the benefits of the town's proximity to the Mississippi River.

Ignatius L. Donnelly, an eastern lawyer, yielded to Nininger's encouragements. Donnelly left Philadelphia, purchased a thousand acres in the township, and moved his family west. Wildly

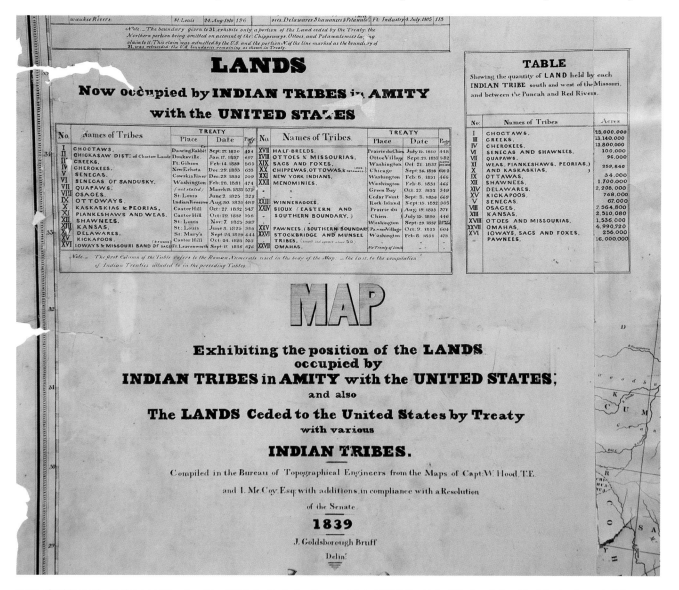

Inset of table on the Indian Land Cession Map, 1839. By the Indian Removal Act of 1830, the United States pledged that the lands west of the Mississippi and Missouri rivers would be laid off into districts, surveyed, and marked so that each tribe would easily know its boundaries.

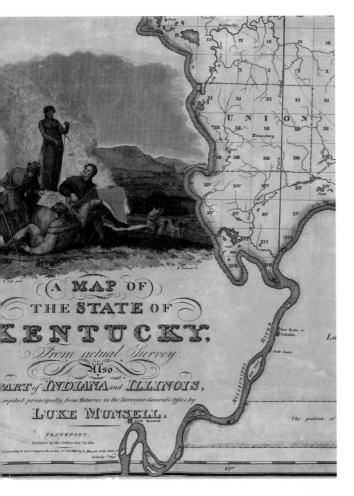

Vignette on map of the state of Kentucky, 1818. This allegorical view offered a Euro-American interpretation of property. Surveyors with their tools measure and divide the land for settlement as the Indian inhabitants leave their homeland. Overlooking the scene, the figure of Justice sanctioned the activities.

SARAH WINNEMUCCA
1844–1891

The Winnemucca family belonged to the Nevada Paiute Indian tribe. Early in Sarah's life her family placed her with Euro-Americans to gain knowledge of white ways. This experience resulted in her becoming fluent in English and Spanish, and with these proficiencies she negotiated between the white world and her Indian nation. First working for the U.S. Army as a translator, then as an activist, educator, and public speaker, she pleaded for the relocation of the Paiutes from the eastern Washington territory to their Nevada homeland. She authored *Life Among the Paiutes: Their Wrongs and Claims* in 1883, with the hope that it would explain the plight of the Paiutes and change land policies. Though many whites supported her cause, the tribe remained separated from their native land.

Sarah Winnemucca, photograph by Norval H. Busey, 1883.

enthusiastic about the town's potential, Donnelly edited the *Nininger Emigrant Aid Journal*, published in English and German, which promoted the town. Nininger and his investors placed broadsides, the *Journal*, and small cards printed in the thousands on trans-Atlantic steamers and in locations throughout the East Coast. Settlers who paid $25 and enlisted in the Emigrant Aid Society received free passage from the East Coast and tools to assist their enterprises in the town. In jest, one of the journals boasted that Nininger would ultimately exceed New York City in population. By 1857 more than five hundred people lived in the township.

NININGER BROADSIDE

Quickly produced notices, such as this, were posted in public venues, and intended to have immediate impact on the viewer. Competition in business and the need to grab the attention of potential customers prompted the use of big letters, bold typefaces, and graphic images to engage the public. With just a passing glance at the broadside, a reader knew that the city of Nininger was accessible by railroad or by steamer. The large, bold type and the word "Emigration" made clear the targeted audience—those considering settlement in a new land. That Nininger had promise was conveyed by its prominently stated location of "Up the Mississippi River" in the "Territory of Minnesota."

City of Nininger broadside, 1857.

During the prosperity of the early 1850s, farmers, merchants, and businesses readily risked buying land in Nininger township. However, the widespread financial panic of 1857 brought chaos. Property values plummeted, settlers lost homes, and investors' assets evaporated. Making matters worse, Nininger failed to attract a railroad line to the town. By 1869, the village had virtually disappeared. Ignatius Donnelly's house remained as a lonely testament to a broken dream.

THE CHINA TRADE

European trade dominated the foreign exchange of the United States in the nineteenth century, especially as the cotton trade expanded rapidly, growing from 16 percent of U.S. trade exports in 1800 to over 50 percent by 1850. But from its beginning, the United States was a global trader and conducted extensive business with Asia as well as Europe and Africa. In the twenty-first century, America's China Trade alone would surpass trade with the European Union and become second only to Canadian trade in total value. Roots of this interchange date back to the Merchant Era.

Indeed, one of the economic consequences of winning the Revolutionary War was winning the freedom from Britain to engage in the lucrative China Trade. Previously the English East India Company provided colonists with desired goods from China, which were taxed at a high rate. Rebellious colonists responded by denouncing the English East India Company and dumping its imported tea in Boston Harbor. After America was a sovereign nation, independent merchants swiftly employed vessels to undertake the economically risky trade with China. With limited resources and little expertise, they engaged the Chinese with their boldness, ambition, and forthrightness. Fierce competition with foreign countries that were already established in the trade with China made Americans resourceful and innovative.

Some market innovations occurred out of desperation. Americans had little hard currency and few articles to sell to the Chinese. Merchants and their ship captains developed circuitous trade routes to locate products of interest to the Chinese: ginseng root from the East Coast, furs from the Pacific Northwest, sandalwood from the Hawaiian Islands, *bêches-de-mer* (sea cucumbers) from islands in the Koro Sea near Fiji, and eventually, opium from Turkey. The trade was unreliable, because after several years of effort the goods might be of little value if merchants encountered a glutted Chinese market.

To shield themselves from risk and the seasonal fluctuations of commodity prices, American merchants stationed partners in Guangzhou (Canton), China. There they purchased Chinese goods when costs were low and warehoused them. Likewise, once these goods came to America, the merchants first stored them, and marketed them only when prices were advantageous. They also convinced Congress not to tax the goods until they were sold. As the trade matured, American merchants beat out their European competition by adopting the new American clippers, the fastest ships available.

JOSÉ ANTONIO NAVARRO
1795–1871

During José Navarro's lifetime, the land on which he lived changed hands three times. Initially it was under Spanish colonial rule, then Mexico. Then it became part of the independent Republic of Texas, and finally the United States. As a rancher, businessman, and merchant, Navarro imported and sold goods on the northern frontier of Mexico. Though he was successful with all these endeavors, his statesmanship in coping with the changing political and economic landscape defined his contribution: that of promoting Texas independence and then statehood, while strongly advocating for Tejanos' (Texans of Mexican American heritage) land and economic rights.

Chinese tea chest, 19th century, paper on wood. George Washington, an avid tea drinker, owned several tea chests. In 1757 he purchased twelve pounds of tea, including best Hyson and best Green. According to Martha Washington's granddaughter, he habitually drank three cups of tea without cream at his breakfast meal.

José Antonio Navarro, photograph, artist unknown, undated.

Spanish silver dollar with Chinese chops, 1805. The Spanish silver dollar, used in America and throughout the world, was often stamped with a Chinese merchant's mark consisting of a punch, a character or a symbol, as seen here. The Chinese preferred silver, although American merchants had less hard currency than other nations with which to trade.

Through their agents, Americans made long-lasting associations with Chinese merchants called the hong, who managed all foreign trade and contact with overseas merchants. The Chinese designated only a select thirteen to oversee foreign trade. In addition, the hong merchants were responsible for the foreigners' behavior and well-being. Americans had to live within a narrowly prescribed and restricted area in the city of Guangzhou, where they also worked.

Wu Bingjian (then known by Americans as Houqua, also spelled Howqua), one of the thirteen Chinese merchants, served as the senior and most respected member of the hong. During his lifetime, the city of Guangzhou functioned as a center of world trade. Still learning the business under his father when the Americans first arrived in 1784, Houqua matured as a merchant under the flourishing commerce with America. Though trading with many nations, Houqua had a special relationship with Americans. Sympathetic to the independent and often precarious nature of their firms, Houqua

View of the Hongs of Canton with the City in the Background, artist unknown, mid-1790s, gouache on paper. This view of the city of Canton, with the residences and businesses of the foreign merchants hugging the water's edge, presented an accurate but romanticized interpretation of the Chinese scene.

here, occupied the parlors of numerous American China traders.

Imported Chinese goods were available in the colonies throughout the eighteenth century. Porcelains, silks, wallpapers, and some lacquerware adorned the homes of the elite. Colonists everywhere consumed Chinese tea. After independence, when Americans conducted their own trade, families of more limited means could afford Chinese products. By 1819, American imports from China had a total value of nearly $8,000,000, surpassing imports from the English East India Company by a half a million dollars.

By the 1830s, many Americans' homes contained some Chinese objects, although their owners may not have realized this fact. One New York City firm, the F. and N.G. Carnes company, capitalized on consumers' capacity to purchase inexpensive goods and imported huge quantities of small items such as fans, baskets, card cases, tablemats, feather dusters, colored papers, floor matting, window blinds, snuff boxes, paper folders, perfumes, toys, buttons, firecrackers, handkerchiefs, shawls, chinaware, sewing thread, silk winders, India ink, umbrellas, combs, canes, beads, and wash basins. In addition, they brought in large quantities of foodstuffs, drugs, and chemicals. Spices, sweetmeats, and preserves, varnishes and paints such as dragon blood, and drugs such as rhubarb made their way into American food pantries and medicine cabinets.

Howqua, **attributed to George Chinnery, oil on canvas, 1830.** Revered by most American traders in China for his advice and monetary assistance in their businesses, Wu Bingjian—Howqua or Houqua—lived a prosperous but precarious life. If events with foreign merchants went awry, officials held him, and his fellow hong merchants, responsible.

advanced loans, forgave their debts, and invested in their businesses. China trader Robert Bennet Forbes noted in his diary: "I consult daily with Hoqua about the general trade & am in his confidence— He is a grand old fellow to us & I wish we were sure of his life for twenty years" (Kerr 1996, 140). As the principal Chinese merchant in Guangzhou, Houqua amassed enormous wealth and during the height of his career was likely the richest man in the world. Mementoes of his largess, impressive porcelains such as the large punch bowl pictured

Chinese hong bowl, hard-paste porcelain, 1785–1795. Americans purchased Chinese earthenware for home use, but this punch bowl was intended for a China trade merchant who understood its meaning. The image depicts the foreign community. Note the American flag. Chinese merchants gave these specialty wares to American merchants as gifts.

AFONG MOY
circa 1817–date of death unknown

In 1834, American China trade merchants, with the assistance of a ship captain, brought Afong Moy, a young Chinese woman, to New York City to promote their imported Chinese goods. Highly publicized as the first Chinese woman to visit America, Afong drew large crowds that viewed her with a selection of Chinese objects in public venues. With bound feet, Afong traveled with a stage manager across the country, informing the public on Chinese customs and foreign goods for those of the middling (now known as the middle) class. The marketing of her distinctive culture along with the related objects significantly increased the sales of these Chinese goods with inexpensive fans, card cases, toys, baskets, and even fireworks, finding their way into American homes.

"Chinese Lady," lithograph by Risso & Browne, 1835.

Chinese fan, paper and ivory, early 19th century. American China trade merchants imported large numbers of fans for they were an affordable fancy good. Available in numerous materials and mounts, the fan was one of China's most appealing and artistic contributions to the Western woman's attire.

When consumers commented on Chinese goods, often it was to express gratitude to those who provided them. In 1833 Elizabeth Latimer thanked her cousin John for the floor matting they received from Canton: "We are much obliged to you….in selecting the matting for us, it is very handsome….Our parlors look very pretty with our new matting on…" Then, as in years to follow, China provided a source for inexpensive goods that consumers incorporated into their daily lives.

opposite, top: **Dragon blood, 19th century.** Exotic in name, dragon blood was merely a dark red resin used in American paints, varnishes, and lacquers. Harvested from the fruit and stem of a prolific Chinese palm tree, the resin was heated, molded into short sticks, then wrapped in palm leaves for export.

opposite, bottom: **Chinese card case, ivory, early 19th century.** Small boxes kept calling cards unblemished. Cases made of bone or ivory, like this one, revealed the marvels of intricate Chinese craftsmanship. Tiny chisels with semicircular blades were used to carve the surfaces.

EXTENDING THE DAY

In the eighteenth century, the sun regulated how people lived their lives. Artisans wove coverlets and cordwainers made shoes only during the daylight hours. Candles provided some light in the evening, but they were costly, their illumination was flickering and undependable, and the light they cast was circumscribed. Betty lamps—consisting of a twisted cloth wick soaked in fats or grease and enclosed in a brass or iron receptacle—were more convenient, but no better.

Innovations that created brighter and more dependable light began in the Merchant Era. These changes modified the very concepts of time, work, and consumption and spurred the market revolution. Industrialists, driven by efficiency, added shifts. The machines need never stop. Shop owners too could increase sales by staying open in the evening. By the mid-nineteenth century, the innovations came into the home; families of even modest means possessed several oil lamps. As the century advanced,

Joseph Russell's New Bedford Oil & Lamp Store, Philadelphia, Pennsylvania, **by Joseph Russell, 1853, watercolor on paper.** Joseph Russell worked as a businessman, but he was also an artist. This depiction of his store provides an unparalleled view of how whale oil was stored, decanted, and sold.

many new forms of lighting emerged and often competed for adoption in different settings.

Whales provided the luxury oil of the early nineteenth century. Liquid at room temperature, it flowed easily up the wicks, burned cleanly, and had little odor. It was the lamp oil of choice. Joseph Russell, proprietor of the New Bedford Oil & Lamp Store on Chestnut Street in Philadelphia, had a great business advantage. His family, of New Bedford, Massachusetts, helped establish the American whaling industry and, by the 1840s, New Bedford had become the world center of whaling. In 1818,

Russell moved to Philadelphia, opened his store, and served as the distributor for his family business— whale oil.

In the 1853 watercolor by his own hand, shown here, Russell replicated the well-appointed interior of the New Bedford Oil & Lamp Store. With precise detail, he illustrated casks of oil and the method of decanting them, as well as whale oil lamps seen on the shelves. Either he, or a well-dressed customer, poured the oil from a pitcher through a funnel into a small container for home use. An African American clerk also appears in the painting, at work, indicative,

Scrimshaw whale panbone, mid-19th century. A portion of the sperm whale's jaw served as a canvas for a whaleman's freehand drawing of the hunt. Seamen engraved images on ivory or bone to fill the monotonous hours at sea; the scrimshaw provides a first-hand record of the whaling event.

Patent model of whale oil lamp, invented by Alonzo Platt, 1836. Platt called this a "union lamp," because the whale oil produced a vivid light by the union of three fuel reservoirs. He recommended his less elaborate lamp be used in cotton factories—but it required a glass beneath to catch sparks before they ignited the cotton.

perhaps, of Russell's support as a Quaker of this segment of the Philadelphia community. Another extant watercolor by Russell shows him outside his store, ready to drive off with his New Bedford Lamp Oil wagon. Although Russell documented his whale oil shop of the 1850s, the use of the oil for lighting purposes waned as Philadelphians turned to other sources of fuel and as the whale population declined due to overhunting.

In addition to the depictions of his whale oil business, Russell also illustrated his home life, revealing, surprisingly, a new fuel source. This watercolor shows a parlor with a table lamp awkwardly fed by a coal gas pipe coming down the wall. Russell and his family lived in Mrs. Amelia Smith's boarding house on Broad and Spruce streets in Philadelphia. In the 1850s, residing in a boarding house was as respectable as owning one's own home; it was even considered fashionable for those of means in American cities. Had Russell owned his own home, he might have continued to use whale oil for lighting his lamps, but because he boarded he had to accept the illumination coal gas provided. It is ironic that he documented the fuel source that would put him out of business, for by 1858 gas lights were considered fashionable and modern. At this time two hundred and fourteen miles of gas pipe connected to over twenty-five thousand households and services in Philadelphia.

In his watercolor of the boarding house parlor, Russell depicted most of the boarders seated by the windows, proximate to natural outdoor daylight. By nightfall they likely pulled their chairs near the sturdy cast-iron and marble table to converse, do handwork, or read by the artificial light of the fan-shaped flame diffused by the shade. The availability of light from gas lamps encouraged sociability, provided possibilities for new types of entertainment, like family games, and extended the day beyond the world of work.

The lighting of factories and retail establishments presented another opportunity for innovation. In 1828, Samuel Slater, with other investors, gained control of a cotton mill on the Merrimack River in New Hampshire and the mill began to prosper. By 1838, the directors planned and later established the factory town of Amoskeag, with streets, commons, housing, and sites for public buildings. Soon after, they built the City Hall, where the social levee (evening party) took place, as noted on the mill ticket shown here.

In 1848, No. 3 was the most recent addition to the mill complex. The levee likely promoted a sense

Mrs. Smith's Boarding House, Philadelphia, by Joseph Russell, 1850s, watercolor on paper. While her boarder sold whale oil, the boarding house owner, Mrs. Smith, used gas. As indoor lighting improved, fundamental concepts of time, work, and leisure changed. Evening hours could be used for reading, playing games, or doing handwork.

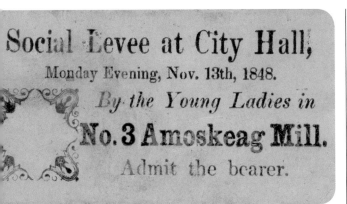

Social levee mill ticket, 1848. Amoskeag was the site of the largest textile mill in New Hampshire. This ticket admitted one of the mill women to a social event likely on the occasion of a changed work schedule with the fall illumination of the mills by oil lamps.

of pride in their new mill. It may also have marked the illumination of the mills by oil lamps in winter. Such an event also encouraged *esprit de corps* during the winter months when operatives spent more than a third of their thirteen-hour days working without the benefit of daylight. At their looms by 6:30 AM, they had thirty-five minutes for a noon break and forty-five minutes for dinner, with dismissal at 7:30 PM.

Factory owners used the steady, bright light of the tinplate Argand lamps filled with whale oil or the weaker light of lamps with lard oil to prolong the workday. Often the light was so dim that the lard oil lamps had to be placed under the looms to illuminate the work. Although the light of oil lamps extended the workday and thus increased output and profitability, open flames created dangerous conditions for workers. Amoskeag Mills fires were common, with the causes rarely recorded. It seems likely that the fires were often sparked by the lamps.

In January 1851, three years after the social levee at City Hall, the directors installed pipes that provided illuminating coal gas in the mills. This more efficient light was a factor in making Amoskeag Mills increasingly profitable. By the end of the nineteenth century, it was the largest cotton textile plant in the world.

Owners of retail establishments also took advantage of the benefits of artificial lighting to promote their goods. Charles Oakford of Philadelphia began his career modestly, as a hat maker in his mother's kitchen. In 1827 he displayed the goods he made in a small twelve- by eight-foot shop. Following the sound advice of a fellow artisan to "… never hold a penny so close to your eyes as to lose sight of a dollar," he then rented a factory and produced hats in earnest (Winslow 1864, 155).

Argand lamp, tinplate over iron, 1800–1840. Factory owners used the bright, steady light of the Argand oil lamp to extend the workday. Swiss Aimé Argand invented the lamp in 1780; eventually its use was widespread in America.

CHARLES OAKFORD & SONS HAT STORE

In 1854 prosperous Philadelphia businessman Charles Oakford installed gas fixtures in his Philadelphia hat store, providing his customers with the latest in lighting to view his merchandise. With stiff competition from other retailers, Oakford also set himself apart by investing in the most current photographic technology to advertise his store. In 1854 photographers William and Frederick Langenheim were the first to use the 3-D stereographic image in Philadelphia. That same year Oakford hired them to complete a 3-D stereographic view of his new store as a promotional item. The novelty of the technique, and initial rarity of the 3-D views, shown here, made the images of Oakford's store a special promotional piece.

By the 1840s, his reputation for making fashionable hats extended beyond his native Philadelphia and soon he added a wholesale business, taking his product nationwide. Recognizing the need for an elegant establishment to display his wares, in 1854 he opened a retail store unlike few others in the city. Ornate pillars terminating in three-light gas fixtures stretched the length of the store. Gas light reflected from the fronts of the glass display cases and onto the marble floor, creating an alluring and dazzling sales environment where goods could be clearly seen and appreciated by the public. A view of his establishment shows hats and other objects on open display, well lit for the shoppers who appreciated the extended store hours to purchase by gas light.

Charles Oakford & Sons hat store, advertising stereograph, by William & Frederick Langenheim, 1854.

Oakford, surely aware of A.T. Stewart's 1848 "Marble Palace" store in New York City, sought to emulate its grandeur. When the Continental Hotel, the palatial new building in Philadelphia, opened six years later, he upgraded again, moving into the grand structure in 1860. Unwilling to leave the elaborate gas light fittings so identified with his establishment behind, he tore them out and reinstalled them in his new store on the

Politics in an Oyster House, **by Richard Caton Woodville, 1848, oil on canvas.** The Baltimore painter subtly depicted new technologies into many of his paintings. On the wall behind the subjects' heads, he recorded the popular form of lighting for public places: a coal gas burner and its piping.

53

first floor of the hotel. Oakford had followed the early advice he received, and never did lose sight of a dollar.

The new culture of artificial gas lighting was epitomized in Richard Caton Woodville's painting of a Baltimore oyster house in 1848. Entitled *Politics in an Oyster House*, the painting depicts a Revolutionary War veteran and a young Mexican War veteran arguing. Generational differences can be seen in the details provided by the artist in the men's clothing. There is a detail of technological change as well—the gas line with the cockspur burner projecting out from the wall effectively separates the two men.

In America, the gas light technologies had begun earlier, in Baltimore, Maryland. It was here too that artists and brothers Rembrandt and Rubens Peale founded the Baltimore Museum in 1814. They drew the public into the museum with innovations and popular entertainment. Always inventive and curious, in 1816 they lit their painting salon with gas, showing visitors a "…ring beset with gems of light…" (Peale 1816). This illumination likely came from a three-gas jet burner, which Rubens developed and had manufactured around 1814. Because of his involvement with this lighting experiment, the brass cockspur burner was named after him. Then with great public acclaim, they went on to establish the first commercial American gas company, lighting Baltimore streets with twenty-eight gas lamps by 1818.

The Peale brothers' technology, as evidenced by the gas pipe seen in the Woodville painting, carried the fuel to the burner with the Peale cockspur, which regulated the amount of gas. With a steady, bright light, the gas provided the illumination of two to three candles. Such light was needed by day and by night in the below-street-level Baltimore oyster cellars, where all elements of society gathered by the light of the cockspur to debate and argue the events of the day.

By the mid-nineteenth century, the market revolution touched the lives of most Americans. The new emphasis on productivity and accountability was evident everywhere. In the factory, clocks measured workers' punctuality and managers ensured a pattern of routinized work. Steamboats and railroads penetrated the interior of the country encouraging new settlements and fostering dramatic shifts in population. In the west, the once wide-open land underwent the furrows of the plow.

The country's population had grown to 23.2 million by 1850 — nearly six times that of 1790. With the admission of Texas into the Union in 1845 and California's admission in 1850, the

Cockspur gas burner, manufactured by Otis Chaffee and Joseph Lyon, 1814. Like their father, Charles Willson Peale, his sons Rubens and Rembrandt were artists with an interest in the natural and scientific world. Rubens developed the gas cockspur burner, which later lit the brothers' Baltimore museum.

United States stretched south to the Mexican border and west to the Pacific Ocean. The country's relatively compact 864 thousand square miles in 1790 had burgeoned to 2,750,000 square miles of farmlands and forests, mountains and plains.

As the Merchant Era ended, the regional and sectional tensions between the North and South were at a fever pitch. A decade later they would lead to the Civil War. In 1850, sixteen of the thirty-one states were slave; fifteen were free. The enslaved, numbering 700,000 in 1790, now totaled over 3.1 million. The interstate slave trade began during the Merchant Era and by its end, the commerce in slaves was the country's most valuable form of investment. The value of slave property totaled at least $3 billion in 1860. To compare, the capital in slaves equaled three times the dollar amount that both the North and the South had financed in manufacturing, seven times that invested in banks, and forty-eight times the amount of U.S. government expenditures in 1860. These economics made proponents of both abolition and continued slavery deeply entrenched in their positions. No resolution would be simple.

The Corporate Era that would follow would bring increased industrialization, more rapid and reliable transportation, and greater productivity in goods and services. Though it led to more trade and less expensive consumer goods, it also helped to make the Civil War the bloodiest conflict in American history. Organized business was replacing the merchant desk and ledgers; international trade expanded well beyond the efforts of the Merchant Era's China traders. The Civil War ultimately brought greater unity. A strong and now unified federal government bound the states together with a stable national currency. The nation was poised for a period of economic growth.

ROBERT CORNELIUS
1809–1893

Robert Cornelius was alert to new technologies. He had acquired his knowledge of silver plating from working in his father's lamp-making business. This, in turn, gave him insight into Daguerre's new photography process. Using this very method, Cornelius took one of the first self-images. Applying the same resourcefulness, he received an 1843 patent on the solar lamp, which burned inexpensive lard oil. The family company would build on this success to become the largest lighting company in America. Always ready for the next innovation, in 1855 Cornelius designed a lamp to burn kerosene, the newly discovered fuel. At that same time another inventor refitted existing lamps to take kerosene, making Cornelius' innovation too late to be profitable.

Robert Cornelius, self-portrait, 1839.

Is This a New Golden Age of Agriculture?

PATRICIA WOERTZ

Patricia Woertz is the chairman of Archer Daniels Midland (ADM), the global crop trading and processing company based in Chicago. Woertz, who became CEO in 2006 after working as an executive at Chevron, argues that American agriculture is increasingly important in the vital task of feeding not only our nation but also the entire world. ADM serves as a critical link between farmers and consumers, turning crops into food ingredients as well as animal feed and fuels like ethanol. It is agriculture, on an enormous scale, an enterprise our Founding Fathers, dare we speculate, would find familiar, unimaginable, but wholly American.

Among the many matters on which our nation's founders had differing points of view was the role and primacy of agriculture in the nascent and fast-evolving American economy. "Agriculture ... is our wisest pursuit, because it will in the end contribute most to real wealth, good morals and happiness," Thomas Jefferson wrote in an August 14, 1787, letter to George Washington (Kaminski 2006). Alexander Hamilton had a broader vision: "[T]he trade of a country which is both manufacturing and Agricultural will be more lucrative and prosperous, than that of a Country which is, merely Agricultural," he wrote in a 1791 report to Congress (Annals of the Second Congress, Appendix 1793).

While history has come down squarely on Hamilton's side, I believe Jefferson would be pleased to see that developments in the years following the Merchant Era didn't supplant the importance of American agriculture; they accelerated and augmented it, driving a level of productivity that has allowed the United States to play a key role in feeding not just itself, but also the world.

Here in the first decades of the twenty-first century, agriculture remains a pillar of the U.S. economy, a defining feature of our national landscape and character, and a primary way by which we help to create a better, more prosperous world. Our vast range of soils and climates enables American farmers to produce an exceptionally broad array of crops and products, from staples like grains, oilseeds, cotton, and cattle to specialty crops such as nuts, berries, citrus fruits, and vegetables of every hue. How did we arrive at this point? And how must U.S. agriculture advance if it is to continue meeting the world's growing need for food, feed, fiber, and fuel?

In 1800, America was a country of farmers. Nearly three-quarters of the free workforce and most slave laborers were employed in farm occupations. It was backbreaking work, delivering subsistence-level results.

The advent of factory-made machinery two decades later spurred mechanization and large-scale farming, and within a decade, commercial wheat and corn belts had begun to develop. That development got a significant boost from the passage in 1862 of the Morrill Act, which established land-grant colleges of agriculture—including Big 10 universities such as Pennsylvania State, the University of Nebraska, and Ohio State—whose research, teaching, and technological developments made significant contributions to national productivity.

For the next seventy years, between 1866 and 1936, newly planted acreage accounted for the vast majority of increased agricultural production. U.S. corn yields remained relatively stable, at about 26 bushels per acre, and wheat yields never exceeded 17 bushels per acre.

Then, over the next five decades—with the introduction of better equipment, fertilizers, breeding techniques, and major investments in export infrastructure and policies that protected farmers' prices and incomes—agriculture experienced a golden age: Wheat yields more than doubled, while corn yields shot up four-fold.

With more crops produced on the same amount of land, the country needed fewer farmers, and a generation moved off the farm into other pursuits, contributing to growth in other sectors. In 1930, 22 percent of the U.S. workforce was employed in agriculture, and agricultural gross domestic product (GDP) represented 7.7 percent of total U.S. GDP. By 2002, the figures were 1.9 percent and 0.7 percent, respectively.

And yet, agriculture remains a vital sector of the U.S. economy, particularly when it comes to exports. In 2012, agricultural exports generated a total economic output of $321 billion, with the United States Department of Agriculture (USDA) noting that every $1 billion of U.S. agricultural exports that year produced 6,577 American jobs throughout the economy.

Importantly, those exports now play an indispensable role in feeding the world. The American Farm Bureau Federation notes that one in three U.S. farm acres is planted for export and that about 23 percent of raw U.S. farm products are exported each year.

Here in 2014, the United States and other major crop-growing nations are working to help meet increasing food demand. By mid-century, global population is expected to reach 9 billion, and with more people desiring better diets, the world must produce as much food in the next forty years as it has in the past 10,000.

If it is to continue playing a leading role serving this demand, U.S. agriculture will need to leverage the innovative capabilities that have brought us the world's most affordable food supply to become even more productive. And throughout the world, we will need to grow more crops and build more capacity to store them safely. We will need more processing facilities to turn them into the products that consumers need and want. We will need more capabilities to move these products around the globe efficiently and cost-effectively. And we will need to do all this with less water and fertilizer, fewer chemicals and less post-harvest and post-consumer waste—issues of importance to the sector and to a public with concerns about the sustainability of agriculture.

I believe that with continued innovation, substantial investments in agricultural infrastructure, and partnerships among all with an interest and a stake in agriculture's future, we can usher in a new golden age of agriculture. The United States, in particular, has a tremendous track record of deploying innovation to improve resource productivity. We have every reason, and every ability, to continue to do so.

Even in this era of phenomenal agricultural productivity, when fewer workers than ever before are needed to plant and harvest crops, there remains in America a reverence for farming. Perhaps it stems from our shared heritage—trace most families' trees, and somewhere you will find a farmer. Perhaps it is something Jefferson expressed

when he wrote in 1795 to Henry Knox, the first U.S. Secretary of War: "Have you become a farmer? Is it not pleasanter than to be shut up within 4 walls and delving eternally with the pen?" Or, perhaps it is just an innate and very human appreciation for the land and the people who feed us.

Indeed, in recent years, a small but growing number of Americans, including young and college-educated students, have turned to farming as a vocation and an avocation. From 2002 to 2007, the number of U.S. farms rose 4 percent—the first recorded increase since 1935—as these new American farmers worked to experience the profound satisfaction of growing and harvesting food.

While Hamilton's vision of a more industrialized America has clearly prevailed over Jefferson's ideal of a nation of citizen-farmers, I believe Jefferson would be highly impressed to see that his ideals still find a home in the heart of America. Both, I am sure, would be proud that the U.S. agricultural economy remains an essential contributor to America's prosperity and to the vitality of the world.

Sources

Annals of the Second Congress, Appendix, 1793. Report to Congress on the Subject of Manufactures, Dec. 5, 1791. http://nationalhumanitiescenter.org/pds/livingrev/politics/text2/hamilton.pdf.

Kaminski, J. P., ed. 2006. *The quotable Jefferson*. Princeton: Princeton University Press, 2006.

How a Family-Owned Business Can Be Sustained

FISK JOHNSON

Herbert Fisk Johnson III, who goes by Fisk, is the fifth-generation family member to run SC Johnson, a global consumer products company that produces the likes of Windex, Glade, OFF, and Saran Wrap. Johnson knows from life experience that competing as a family-run business for so many decades has required his family to operate in ways that differ from most other companies. He joined the company, which is still owned by his family, in 1987 after earning bachelor and advanced degrees from Cornell University in Engineering, Physics, Chemistry, Marketing, and Finance. Many businesses in America start out as family companies; SC Johnson is one of the very few to grow so big and survive so long.

American history is full of businesses named for the people who founded them. Levi Strauss & Co., Woolworth's, Ford Motor Company, Eastman Kodak. They all shared traits with the family businesses that fueled the American economy as far back as the Merchant Era: they seized the opportunity to create a livelihood with their unique skills, and as needs, technologies, and society evolved, they continued to expand their offerings and their reach.

My great-great-grandfather, Samuel Curtis Johnson, was just such an entrepreneur. In 1886, after several failed business ventures, he bought a parquet flooring business in Racine, Wisconsin. It was a successful small business, but Samuel saw a bigger opportunity when customers asked for products to care for their new floors. By 1898, floor care sales had exceeded those of the flooring itself.

And so it goes with family companies—an initial idea proves successful and, in the companies that thrive, each generation adds something new to the enterprise. But over time, while some family companies evolve, others are sold or fade away. Few multigeneration family companies are still led by the families who created them. At SC Johnson, we remain a family-based business, just as we were in our inception during the Merchant Era. As we've scaled to a global consumer products company, we've kept the core philosophy that guides us.

So what is it that makes a family company endure? There are many answers to that question. For SC Johnson, we have long been grounded in a commitment to the greater good. It is deeply engrained in our company's values. At times, it has led to bold decisions that have set our company apart.

Take my great-grandfather, for example, our second-generation leader. Herbert F. Johnson Sr. championed international expansion of the floor care company his father started. But his most lasting contributions are the remarkable decisions he made for people.

Until the Fair Labor Standards Act of 1938, there was no national minimum wage or protection for workers. Industrial growth, for many, meant poor working conditions and unfair pay. Yet in 1917, Herbert launched eight-hour workdays, as well as one of the first profit-sharing programs in the country. That first year, 193 employees shared $31,250. Today, tens of millions of dollars in profit shares are distributed. The company is blessed with exceptional commitment and employee engagement.

My grandfather, H.F. Johnson Jr., carried on this commitment to the greater good. He liked to say that we couldn't have a healthy environment inside the company unless we had a healthy environment outside its walls. We wouldn't be able to hire good people for the business, for example, unless there was a good community for them to live in. H.F.'s vision stretched far beyond our company and its neighborhoods.

Just one example was the 1964/65 World's Fair. Hosted in New York, the fair aimed to shine a spotlight on the wonder of American industry. IBM would demonstrate the magic of computers. Bell System would debut a "picturephone." RCA would give visitors a chance to actually see themselves live and in color on a TV screen.

H.F. wanted SC Johnson to participate, but not as just another American business in an industrial display hall. This was the early 1960s. The American President had been assassinated, issues in Vietnam were escalating, and the battle for civil rights was raging throughout America. Against that backdrop of upheaval, H.F. wanted to make a documentary film about peace, understanding, and the joy of being alive.

I'm not sure a project like this would have proceeded in a public company. Our own management team was aghast at the thought of spending a portion of that year's marketing budget on a show that had nothing to do with the company's products. But when the film *To Be Alive!* premiered at the fair, it was showered with praise. H.F.'s bold film transcended business, it transcended the fair, and it brought to life a vision of greater good for the entire world. It became one of the most talked-about exhibits at the fair, won an Academy Award, and lives on today as an example of corporate social responsibility.

My father, Sam Johnson, also put doing what was right above doing what was expected. In 1975, he shocked the chemical industry by banning chlorofluorocarbons, or CFCs, from the company's aerosol products worldwide–despite the fact that the research on the danger of CFCs was still incomplete.

Other companies accused Dad of being irresponsible and trying to wreck the chemical industry. But Dad believed it was the right thing to do, and three years later, the U.S. government agreed, banning CFCs as an environmental hazard. This initial bold decision laid the groundwork for decades of leadership around environmental issues, transparency, and sustainable chemistry.

I faced a similar challenge, but on a lesser scale, shortly after SC Johnson acquired Saran Wrap from Dow Chemical in 1998. At the time, concerns were growing over the use of polyvinyl chloride (PVC) because it could emit toxic chemicals if it was incinerated in the waste stream. Saran Wrap did not contain PVC, but PVDC was chemically similar enough that it was very likely to cause the same issue. The inclusion of PVDC in Saran Wrap gave it an advantage over competing products. We faced a decision–continue on as is, or remove it as an ingredient and potentially be disadvantaged in the marketplace. Our RD&E team worked hard to find an alternative that would maintain the product's competitive advantages, but no viable solution presented itself. Even though the scientific evidence was less than tangible, we decided that the right thing was to take extra caution and change the formulation, resulting in a less effective product and ultimately resulting in a significant loss of business. But I sleep a whole lot better at night as a result.

One of the most difficult challenges we faced was in the 1980s. Dad stuck to his guns again, when the company was attacked on all sides for continuing to operate in South Africa, where we had been for two decades. At the time, our business there was less than 1 percent of the company's consolidated worldwide sales, so it would have been easy to leave. And the pressure was intense to do so, there were large-scale public campaigns encouraging U.S. multinationals to leave South Africa as a protest against the harsh apartheid regime. I'll never forget walking out of a class when I was at Cornell University and seeing Dad being burned in effigy in a student protest of apartheid.

Many people felt the only right choice for a company was to leave South Africa, but Dad chose to stay. He believed he could have a greater impact by fighting for change from within, not from afar. He would never consider abandoning the loyal people who worked there.

While other companies pulled out, Dad withstood the criticism and focused on being an agent of change. The company continued to operate as a model of integration. It ensured equal treatment for all workers, and provided health care, educational aid, housing, wages, and community support far above that of local corporations. Over time, these types of steadfast examples from multinational corporations helped shape local business and government opinions.

While bold, my father's actions were simply a continuation of a long-held tradition that has been passed along to our entire family. My great-grandfather believed "that every place where we operate should become a better place because we are there." As a family company, every choice goes through a filter that crystallizes even the toughest decisions. We ask ourselves: Will doing this make life better and serve the greater good for generations to come? That opportunity is a responsibility we never take lightly. As a family-owned company, we have a greater ability to look long-term and do what is right for the greater good and for the company, rather than worry about next quarter's earnings.

My dad used to say that as leaders we should not worry about whether we live up to the expectations of our fathers, but rather if we as fathers live up to the expectations of our children.

As the leader of a fifth-generation family company, and as a father, I can tell you: This, above all else, is the boldest decision an American business can make.

Standard Oil refinery, Cleveland, Ohio, 1890. Standard Oil's reorganization as a corporation and the aggressive acquisition of competitors in the 1870s set a new tone for business in America.

The Corporate Era

1860s–1930s

Edison printing stock ticker, 1873. Thomas Edison used money that the Gold and Stock Telegraph Company paid him for his improvements to the stock ticker to set up an independent research laboratory in Newark, New Jersey. The concept of a dedicated lab was an important invention in itself.

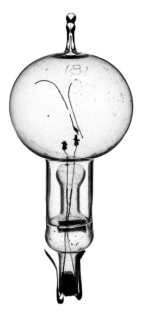

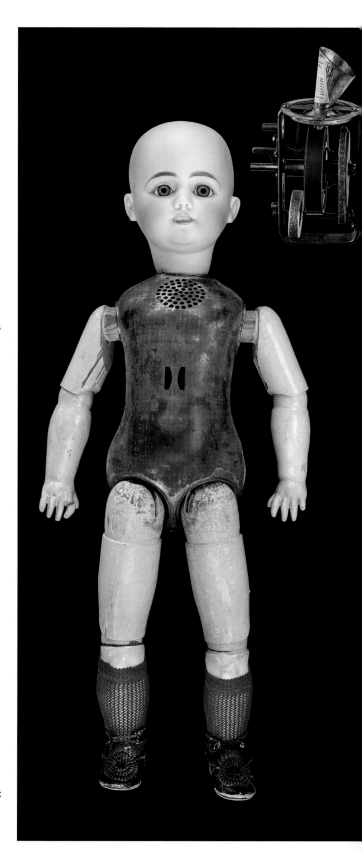

above: **Edison light bulb, 1879.** The first public demonstration of electric lighting took place at Edison's Menlo Park, New Jersey, lab on New Year's Eve, 1879. Edison's improvements on light bulbs made electrical lighting practical but to make it useful, Edison had to invent an electrical distribution system as well.

right: **Edison talking doll, 1890.** In order to make a talking toy, the mechanical playback system (a wind-up phonograph) had to be shrunk small enough to fit into the body of the doll. Each cylinder was individually recorded. Heavy, expensive, and with poor audio quality, the doll was a marketplace failure.

THE CORPORATE ERA was a time of tumult and change for the United States of America. Production efficiency and individual consumption began replacing the founding American values of individualism and egalitarianism. Factories with tall chimneys belching smoke, office buildings filled with hundreds of clerks filing papers and clicking away on typewriters, and department stores packed with goods became new symbols of success and national identity. By the 1920s half of all Americans had left rural farms and moved to cities, where they could engage fully in the risks and rewards of American business.

A hero-inventor, Thomas Alva Edison (1847–1931) became one of the icons of the Corporate Era and his life symbolized the times. With a penchant for self-promotion and 1,093 patents for ideas as varied as electrical lighting to motion pictures, Edison was known for success. His career however, like many in business, began with a business failure. In 1869 he patented a vote recorder based on telegraphic principles. Trying to sell the invention to the United States Congress, he was turned down. Preferring to maintain the politicking opportunities that a slow roll-call vote offered, the committee chair at the time proclaimed, "If there is any invention on earth that we don't want down here, that is it" (Rutgers University). Edison learned a lifetime lesson, "Never waste time inventing things that people would not want to buy" (New England Historical Society).

Undeterred, Edison moved on with a string of inventions that did succeed and smart business choices that made him wealthy. In 1869 he moved from Boston to New York City, where his work in printing telegraphy was highly valued by both the financial and communication industries.

His focus would soon turn away from telegraphy though, and towards electrical systems, what some now call the second Industrial Revolution. Edison built his first full-scale power station on Pearl Street in New York City, in part because that location allowed him to impress investors on nearby Wall Street but also to take on his main competitor— established gas lighting systems.

In 1877 Edison's invention of the tinfoil phonograph launched the field of recorded sound and playback. He did little with the idea until 1887 when, after Alexander Graham Bell, Chichester Alexander Bell, and Charles Sumner Tainter had developed wax cylinder sound recording, Edison tried to profit from his earlier invention of the phonograph and produced a doll that talked. It was a disaster in the marketplace. The doll was expensive, it was also heavy, and the quality of its all-important voice was, in a word, poor. Very few sold. Edison himself called them "his little monsters." The demise of his doll did not deter Edison. Optimism is an American trait, and Americans tend to overlook failure. Ideas were free—and, for Edison, plentiful.

In general, American capitalism encouraged investors (both large and small) to take risks and potential failure in exchange for financial rewards. The division of profits from a success, however, was not obvious: how much of a return was appropriate to investors who had risked their fortunes? How much to managers whose wits had moved companies forward? How much to workers from whose sweat the products had come?

By the late 1800s the influence once held by merchants and traders began to fade and the power of industrial entrepreneurs like steel magnate Andrew Carnegie, oil mogul John D. Rockefeller, and meatpacking king Philip Armour began to dominate American life. Some Americans worried that a new aristocracy—of industrialists and robber barons—was being enabled. New machines also brought change. Innovations in communication (telegraph and telephone), lighting (gas then electricity), and transportation (steamships and railroads) affected everyday life and made business more efficient. It was a new age.

During the Corporate Era, change extended to the general population as well as to machines and systems. Migrants from around the country and immigrants from around the world flocked to industrial centers like New York, Pittsburgh, Cincinnati, Detroit, and Chicago, seeking opportunity. Skilled artisans, like tinsmiths and coopers,

THE "BRAINS"

THAT ACHIEVED THE TAMMANY VICTORY AT THE ROCHESTER DEMOCRATIC CONVENTION.

"The 'Brains,'" by Thomas Nast, *Harper's Weekly*, October 21, 1871. Thomas Nast's political cartoons lampooning the corrupt practices of New York politician William "Boss" Tweed fed Americans' anxieties about any alliances between government and capital. Tweed was eventually sent to jail.

disappeared, replaced by laborers taking semi-skilled jobs in sprawling factories. A growing middle class, including female office workers and saleswomen, made more money than ever before and they spent more of it on an increasing variety of manufactured goods. The luxuries of yesterday became the necessities of today as the standard of living of urban dwellers ticked ever upward.

The economies of scale, the division of labor, and the growth of national markets brought many positive changes, but the concentration of wealth and power also created a crisis of control and threatened democracy itself. For some people the Corporate Era provided great affluence, for others poverty. For most, it was a time of a growing middle class, one that was earning and consuming more.

There was no escape from the turbulence of the Corporate Era; massive changes confronted both producers and consumers wherever one turned. From the rise of the efficiency of big business with its dramatic effect on competition to the growth of markets within the country and around the world, change was drastic.

BIG BUSINESS

In the second half of the 1800s, industrialists and financiers raised large amounts of capital–money– and made operations ever bigger as they sought to build upon the pioneering industrial successes made possible through economies of scale, division of labor, and labor-saving equipment. With an increase in efficiency and improved transportation, businesses became more competitive, expanding to national and international markets.

The Singer Manufacturing Company, formed in 1851 as I.M. Singer & Co., then renamed in 1853, was an early example of American big business. The sewing machine, produced in giant factories, was a breakthrough technology—what today is called "a disruptive technology." The use of the sewing machines in factory production with a division of labor replaced the old practice of sewing by hand in small quantities. Also used in homes, sewing

machines increased productivity significantly. Clever marketing expanded sales and a vast new office workforce helped Singer's management maintain control and profitability of a widely distributed operation. By 1900 Singer was a model of what would become the modern multinational corporation.

As is often the case with disruptive technology, the innovation of machine sewing was not an immediate economic success. Neither individuals nor companies were rushing out to buy the expensive new machines. One of the problems that hindered the development of the sewing-machine industry was the fact that several inventors had competing patents, what legal scholars refer to as a patent thicket. In the case of the sewing machine there was no single inventor with an overriding patent who could exercise control. Instead there were many small component overlapping patents on individual parts that were owned by different people. This fragmentation of intellectual ownership of the sewing machine led to costly litigation that at first restricted the growth of the sewing-machine industry. Orlando Brunson Potter, a lawyer and president of the Grover and Baker Sewing Machine Company, solved the problem by negotiating a patent pool in 1856 (the first such combination in American history). The pool included three sewing-machine manufacturers: Singer, Grover and Baker, and Wheeler and Wilson, plus Elias Howe, a sewing-machine inventor who had a claim on the original sewing-machine concept but never set up a manufacturing company. Individuals like Howe who patent an idea but do not fully develop it or bring it to market and then litigate those that do manufacture are often now called patent trolls. The sewing-machine patent pool turned out to be a success without the interaction of the courts. Their hard-fought agreement: any company that made sewing machines would pay a $15 licensing fee per machine to the pool. From that pool, a $5 royalty would be paid to Howe for each machine sold in the United States, with $1 for each machine exported. A portion of the remaining monies would be used

Singer advertisement, 1901. One of the first American companies to aggressively seek worldwide sales, Singer was financially well served by its marketing strategy. Managing such a far-flung empire with factories and sales offices in many nations, however, was an administrative nightmare that required structured reporting and a mountain of paperwork.

for a legal defense fund and the remainder would be divided among the four patent pool members.

Inventor Isaac Merritt Singer (1811–1875) had an active role and financial interest in the Singer Manufacturing Company, but it was Edward Clark, a lawyer, who served as its first president and whose leadership and vision established a long-term direction for the business. Clark expanded sewing-machine use from industrial applications to home use and he promoted the sewing machine as a means to improve one's life–"the great civilizer." Clark responded to the 1856 recession by instituting installment buying. For $5 dollars down and $3 a month, a person could rent to purchase a machine. To prevent the rise of a competing second-hand sewing-machine market, Clark instructed sales agents to accept Singer trade-ins for new machines and to destroy the old machines. The company followed a conservative financial policy, reinvesting roller-coaster profits into low-yielding but safe

Singer sewing machine patent model, 1851. Guaranteed in the Constitution, patents give inventors limited monopoly rights. Isaac Singer patented improvements to the sewing machine that gave him the exclusive right on his idea for seventeen years, but did not cover the fundamental concept of a sewing machine.

securities. The resulting solvency meant Singer could ride out cyclical downturns like Black Friday in 1873, the Panic of 1907, the Great Depression, and the two World Wars. Equally important was Clark's choice to enter foreign markets. In 1864 when many manufacturers were suffering the disruptions from the American Civil War, Singer's foreign exports accounted for 40 percent of total sales with offices in South America, Europe, and Russia.

Frederick Gilbert Bourne was president of Singer from 1889 to 1905, and it was Bourne who began to centralize control of the Singer empire in New York City. He instituted bureaucratic policies and forms designed to give upper management control over day-to-day operations and monitor cash flow. Agents sent weekly reports to headquarters and a standard payment schedule was begun. Businesses, like Singer, became large and impersonal, and extension of credit became difficult as owners and managers ceased to know their clients personally. Credit reporting agencies like R.G. Dun & Company solved the problem by putting investigators in the field who compiled often very personal information on the net worth, duration of business, source of wealth, and character and reputation of individuals and firms. Reading the reports gave manufacturers and distributors the confidence to accept orders and extend credit to businesses in faraway cities or states that they did not know.

By the beginning of the twentieth century many factories around the country, from can making to clock manufacturers, adopted the principles of mass production—large volume, installation of specialized machinery to eliminate human labor, interchangeable parts, and the use of low-skilled workers. Perhaps no company is more associated with mass production than the Ford Motor Company. Established in Detroit, Michigan in 1903, the Ford Motor Company was the third attempt by Henry Ford (1863–1947) at automobile manufacture. At first the company sold a number of medium-priced cars made in a traditional factory. In 1908, Ford introduced the Model T, a lightweight car for the masses. Initially the car cost $850 (about an

Underwood #5 typewriter, 1914. As companies grew bigger and their distribution networks more widely dispersed, they hired many office workers to process correspondence, memos, and other sorts of paperwork. With the introduction of typewriters for correspondence, women, who had little chance of advancement, replaced men as clerks.

average American's yearly salary), but production efficiencies allowed the price to drop to $360 by 1915. The improvements came in the form of more interchangeable parts, the use of specialized machines, a high degree of division of labor, continuous flow, and increased speed. By 1913 engineers began using a moving assembly line for some components of the car and the modern car factory was born.

An assembly line increased speed and efficiency as it reduced the need for skilled labor. Factory work, however, became repetitive and boring. Ford and other manufacturers, from steel mills to tire plants, turned to the massive number of immigrants entering the United States and to African Americans leaving the rural South to staff their factories. Due to the routinization and frenzied pace of work and because job opportunities abounded, labor turnover in the Ford Highland Park plant became very high. In 1913 the turnover rate reached 380 percent. Throughout the 1910s when profits were substantial, Henry Ford's vision of capitalism was, in a word, optimistic: high wages, high production

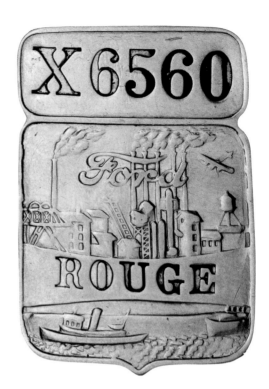

above: **River Rouge worker ID (top) and tool check (bottom), about 1940s.** Workers at the Ford Motor Company River Rouge plant in Detroit, Michigan, were issued ID badges to help managers control a large workforce that they did not personally know. Tool checks were retained when workers checked out tools from a central tool room.

left: **Time clock, 1915.** Factory work and assembly line production meant employees had to work the same hours. To enforce the rules, managers made workers punch time cards. Grudgingly, workers went along with the regimentation of work and the loss of self-determination. Few companies had managers punch in.

of low-cost goods, and high consumption. He argued that workers should be paid enough so that they could consume the products that factories were turning out.

In 1914 the company sought to ameliorate the employee turnover problem and announced the unheard-of plan of a $5 day—nearly doubling the $2.34 minimum Ford wage of 1913. As a result, the company received extensive free publicity as well as long lines of job applicants. In reality, few workers actually got the high pay, and those who did had to follow strict rules extending to their personal life. The company went so far as to send nurses to check on their home life. As the U.S. economy went through a down cycle and Ford's big competitor, General Motors, introduced annual model changes and automobile financing, Ford's market share tumbled from 55 percent in 1921 to 30 percent in 1926. With car sales dropping, Ford's attitudes towards workers became harsh. His assembly line workers were forbidden from talking, whistling, or sitting down, and in-house spies reported fellow workers who violated the rules.

In the early twentieth century, both industrial and office work became drearier as management consultants like Frederick Winslow Taylor, and Frank and Lillian Gilbreth used motion studies to analyze workers' movements in an effort to separate the thinking and doing part of work. As managers sought to increase efficiency, workers feared being turned into machines.

For many manufacturers, from sewing machines to agriculture equipment, the success of new production systems and techniques encouraged them to seek new markets expanding nationally and soon internationally. Mass production required mass consumption.

LILLIAN GILBRETH
1878–1972

Psychologist, industrial engineer, and mother of twelve, Lillian Gilbreth was a barrier breaker earning a PhD from Brown University in industrial psychology in a time when few women attended college. She began her career working with her husband Frank to develop the field of scientific management, doing time-motion studies in factories to improve efficiency and relieve worker fatigue. Using motion study photography, the Gilbreths searched for the "the one best way" to do the job. She also redesigned the layout of domestic kitchens into triangular work spaces, increasing efficiency at home, and worked with large companies, like Macy's, on office management.

Frank and Lillian Gilbreth and their children in their car "Foolish Carriage," about 1915.

ALFRED PRITCHARD SLOAN
1875–1966

Becoming the president and chief executive officer of General Motors in 1923, Alfred Sloan reimagined the management of the business and the marketing of the automobile. He centralized financial planning, decentralizing production, and used techniques like an annual model change, consumer financing through GMAC, and the idea of a car for "every purse and purpose" to make business more efficient and change consumer attitudes about the automobile as something to be upgraded yearly. Over his long tenure at GM, he not only grew his company to surpass rival Ford but also created a template for the modern corporation.

Alfred Sloan, 1924.

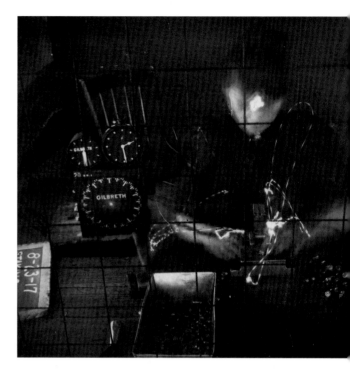

Gilbreth motion study, "staking buttons," 1917. Frank and Lillian Gilbreth used time-lapse photography to record workers' movements. From these photos, they built wire models of the techniques of the very best workers. They reasoned that by studying highly productive workers others could be trained to be equally fast.

FOREIGN TRADE

Foreign policy was another significant factor in the rise of industry in the United States. The country used military and political power to dominate world trade and increase American national wealth. Working together, big business and government controlled international sources of raw materials, secured international markets for selling finished goods, protected domestic interests through tariffs and "Made in America" campaigns, and protected (or expropriated) intellectual property through laws and regulations.

Kentucky Senator Henry Clay (1777–1852) was a staunch supporter of protective tariffs. He argued that tariffs would protect infant American manufacturing firms and the proceeds would finance infrastructure construction. In 1844 Clay ran for president and promoted himself as the "Champion of the Protective Tariff." Over forty years later,

Benjamin Harrison (1833–1901) ran for president on a similar campaign that promoted "America First." Elected president in 1888, Harrison signed into law in 1890 the highly protective McKinley Tariff. It increased the duties on wool, woolen manufactures, tin plate, barley, and other agricultural products and eliminated the duty on raw sugar. The bill also had a reciprocity feature that provided for the remission of duty on certain products from those countries that removed duties on products imported from the United States. The McKinley Tariff was a major factor leading to the Panic of 1893 and the deep and lengthy depression that followed. Forty years later in 1930 one of the highest tariffs in U.S. history, the Smoot–Hawley Act, was passed just as the United States and other countries plunged into depression. The legislation dramatically raised tariff levels on imported goods in an effort to protect American production. Foreign trading partners, in turn, retaliated with higher tariffs of their own, and American trade was reduced by about half. The actions contributed to the depth and length of the global depression.

White House dessert plate, 1892. Elected after running on an "America First" campaign, President Benjamin Harrison purchased White House china that featured a decorative pattern of North American icons—corn and goldenrod. Ironically, the plates were manufactured by Tressemanes and Vogt, a porcelain maker, in Limoges, France.

Chinese Coca-Cola ad, about 1935. Part of the appeal of Coca-Cola is cultural not taste. This ad promotes the notion that consuming Coca-Cola is sophisticated and modern. Close inspection of this "Shanghai lady" style advertising poster reveals many symbols (hair style, plate of food, shoes, et cetera) of modern Western life.

In the 1890s a growing number of American companies sought to increase sales by expanding into foreign markets, and some of them were very successful in their efforts. In 1902, foreign sales of the McCormick Reaper Company reached $4.3 million, nearly 20 percent of total company sales. Singer's foreign sales went from 52 percent in 1879 to 83 percent in 1912. In 1906 Coca-Cola entered international markets, setting up a bottling plant in Cuba. The company then expanded, creating a Foreign Department in 1926, and by 1939 had 41 foreign bottling plants, making international markets an important aspect of company growth.

RISE OF ADVERTISING AGENCIES

The advertising business came of age between 1870 and 1920. Advertising matured along with mass production and the development of national markets, and advertising agencies developed selling strategies, like branding and the launching of national campaigns, to ensure a steady demand for new products and a growing culture of mass consumption. A new breed of advertising professionals created full-service businesses—which not only designed ads but also placed them. They staked their success on legitimate business practices implementing transparent billing systems, forming associations to share information, and promoting themselves with slogans like "Advertise Judiciously." Professional associations allowed advertisers to share information, standardized billing, and develop a code of ethics that would guide agencies and quell suspicions about disreputable business practices. The American Association of Advertising Agencies (AAAA), formed in 1917 by five regional associations, worked at a national level to raise standards as well as advocate on behalf of the profession. A year later, the AAAA (which later became known as the 4A's) created a set of standards for service that facilitated business between agencies, publishers, and those paying for advertising.

Board of directors for the American Association of Advertising Agencies gather at the Lantern Club in Boston, Massachusetts, October 1919.

Protecting foreign sources of raw materials for American consumers and companies was as important as retaining foreign markets. Beginning in the mid-1800s, the sugar industry dominated royal Hawaiian politics. American tariffs, free trade agreements, and other rules impacted Hawaiian sugar competitiveness. Seeking annexation in 1893, American business leaders, with the visible support of soldiers from the USS *Boston* anchored in Honolulu harbor, deposed Hawaiian Queen Lili'uokalani. In 1898 the United States annexed Hawaii as a territory.

Bananas were also of paramount importance. By 1910 bananas were one of the most widely eaten fruits in the United States, yet all were imported. Refrigerated banana boats were a key link in bringing perishable bananas to the U.S. market. Around 1900, American shipping companies, like United Fruit and Standard Fruit, vertically integrated (expanding downward into production and upward into distribution) to dominate the banana trade. They bought vast tracts of Central American lands and constructed infrastructure in order to grow and ship bananas. The American government helped by bringing stability to the politically turbulent "banana republics" of Central America and the Caribbean. In a show of world force, in 1907 President Theodore Roosevelt (1858–1919) sent sixteen battleships and support vessels to circumnavigate the globe on a yearlong "good will" tour.

COMPETITION

Technological, financial, and organizational innovation, coupled with the rise of national distribution, lowered costs of production and goods but created an environment of cutthroat competition. Consumers, of course, liked the low prices that large factories made possible. By the 1890s, however, the public was nervous about the power and influence of the owners and managers of such powerful companies as Standard Oil and Carnegie Steel. Americans began to debate the virtues and detriments of free enterprise and the need for

Political cartoon, "Next!," by Udo Keppler, *Puck* magazine, 1904. By the end of the nineteenth century, a growing number of people believed that large corporations like Standard Oil were dangerous monopolies destroying the American tradition of independent entrepreneurs. In this cartoon the monster, Standard Oil, has most of government and industry in its tentacles and is threatening the White House.

government regulation. While the Sherman Antitrust Act was passed in 1890, the number of mergers continued to increase. Three products and services— kerosene for lamps, steel for railroads, and telephone communication—provide insight into the different ways that industry tried to control competition and how the public and government reacted.

The early American oil industry was chaotic. Production of kerosene, patented in 1854 as a cheap alternative to whale oil for lamps, quickly outran demand. By 1861 prices dropped from $10 a barrel to $0.50. John D. Rockefeller (1839–1937) entered the refinery business in 1863. He hated "unbridled competition," and in 1870 he formed the Standard Oil Company to control rampant overproduction

of kerosene. Rockefeller fought to control the kerosene market by expanding horizontally, buying up his competition and developing customers at home and abroad. By the end of 1878 Standard Oil controlled slightly more than 90 percent of the refinery capacity of the United States but stayed out of drilling and pumping. And because Standard Oil was such a big customer of the railroads (the predominant means of transporting oil), Rockefeller could demand kickbacks (a legal maneuver at the time) on transportation rates. He bought up many of his refinery competitors (often through mergers) and those that would not sell he ran out of business with ruthless price cutting made possible with the lower cost of operations.

Andrew Carnegie followed a substantially different path as he built the nation's largest steel empire. He expanded his enterprise not horizontally as Rockefeller had, but vertically—from raw materials like iron ore and coal to finished goods such as steel rails. Like Rockefeller, Carnegie pumped profits into new technology, in his case Bessemer furnaces and new rolling mills. By investing in technology, Carnegie increased his capital expenses significantly, but in turn raised quality and capacity while lowering the time needed to produce a product and ultimately his unit costs. Nearly half a century before Henry Ford revolutionized the auto industry, Andrew Carnegie brought mass production to the steel industry.

Carnegie also instituted tight accounting practices and slashed labor costs. He explained: "I insisted upon such a system of weighing and accounting … as would enable us to know what our cost was for each process and especially what each man was

Oil refinery in Cleveland, Ohio, the Standard Oil Company, 1890. John D. Rockefeller created a horizontal monopoly controlling most U.S. petroleum refining in the 1880s. He led Standard Oil aggressively, acquired competitors, and invested in new technology in an effort to halt price wars and bring order to a very competitive market.

John Rockefeller and son, 1921. In 1902, literary muckraker Ida Minerva Tarbell wrote an exposé series in *McClure's Magazine* where she claimed that John Rockefeller's Standard Oil was destroying small independent oil refiners. Her dramatic prose was not completely accurate, but it helped motivate the federal government to pass antitrust legislation.

doing, who saved material, who wasted it, and who produced the best results" (Nasaw 2006). A business visionary, Carnegie used basic cost accounting—running the business by personally looking at the account books. He also viewed labor as a cost like limestone. Carnegie believed in the moral philosophy of British intellectual Herbert Spencer (survival of the fittest) and market forces. Carnegie took the unusual step of suggesting that wages should go up when the market boomed and wages should go down when the market was depressed. In December of 1887, Carnegie put this pragmatic industrial philosophy into play as he negotiated a new contract with the Knights of Labor. He locked out the workers at his Edgar Thomson mill in Braddock, Pennsylvania, and offered the workers a contract featuring a sliding pay scale (fixing wages to steel prices), a 10 percent wage cut, and a return, for skilled workers, from an eight-hour day to a twelve-hour day (two shifts instead of three). The workers said no. As the strike progressed, Carnegie brought in Pinkerton guards and threatened to hire strike breakers. In April the unskilled workers returned, followed soon by enough skilled workers to run the mill. With the strike over, the Knights of Labor were banned and the mill run as a nonunion operation on Carnegie's terms.

Five years later during the bloody Homestead, Pennsylvania, strike of 1892, events followed a different arc but led to a similar conclusion. Carnegie went on vacation and let his lieutenant Henry Clay Frick bring in Pinkerton guards and scab workers to break the strike. Both workers and Pinkertons resorted to violence and the Pennsylvania state militia was brought in to quell the protest. In the end, the workers lost, and the union was tossed out.

Carnegie always sought to expand his market share, underselling competitors, brutalizing labor, and keeping wages and salaries low. At times he was willing to sacrifice profits to gain market share, even keeping up steel production during economic downturns such as the depression of the 1870s. Building the Edgar Thomson Works, a state-of-the-art steel mill near Pittsburgh, Pennsylvania, and bringing on capable managers like Henry Clay Frick and Charles Schwab, Carnegie Steel stayed well ahead of the competition. Switching production at the Homestead mill from rail to plate and structural iron in the late 1900s kept the company in front of consumers' demands. Andrew Carnegie, pushed his steel mills hard. He often pitted department against department, mill against mill, and so on in a never-ending race to increase productivity. Carnegie, like other American capitalists, installed innovative new technology to diminish not preserve labor and lower material costs. Fiercely competitive, Carnegie continually sought to increase efficiency with little regard for traditional practice.

ANDREW CARNEGIE
1835–1919

Born in Scotland, Andrew Carnegie immigrated with his parents to Pittsburgh in 1848 and found employment as a telegraph messenger. While working for Tom Scott, superintendent of the Pennsylvania Railroad, Carnegie learned the techniques of "crony capitalism." After the Civil War, Carnegie worked as a bond salesman parlaying his commissions into a lucrative investment portfolio. Building the nation's largest steel company, the Carnegie Steel Company, he invested in new technology, instituted tight accounting, and ruthlessly slashed labor costs. In his book *The Gospel of Wealth*, Carnegie argued the common good is served by allowing men like him to accumulate and retain huge fortunes. The more wealth that landed in their wise hands, the more that could be given away.

Andrew Carnegie, about 1905.

Bell telephone, 1876. Alexander Graham Bell used his experimental telephone at the Philadelphia Centennial Exhibition to demonstrate the invention of telephony. Unlike telegraphy it allowed electrical communication over wires without special skills. By 1900 telephones had moved from a revolutionary oddity to an everyday appliance.

Theodore Newton Vail (1845–1920), president of the American Telephone and Telegraph (AT&T) from 1885 to 1889 and again from 1907 to 1919, managed competition by building one of the biggest monopolies in U.S. history. At first the company used its Alexander Graham Bell patents to protect itself from competition. But when Bell's second patent, which covered the fundamental principles of telephony, expired in 1894, competitors jumped in. By 1904 there were over 6,000 independent telephone companies in the United States and interconnection was a problem. In 1907 Vail argued that a legally sanctioned monopoly was beneficial to the country by providing "One Policy, One System, Universal Service." The federal government agreed, and until 1984, AT&T operated as a "natural monopoly," where set-up and infrastructure costs make a single company more efficient. The company also continued to protect itself through innovation by creating the AT&T Bell Laboratories, once the nation's most famous science-based fundamental research facility.

The public acceptance of AT&T as a benign monopoly was a fairly isolated case. By the turn of the century, the American public had begun to fear that large corporations might fix prices and cheat

27   E Y   1921

Democracy

"—of the people, by the people, for the people"

People of every walk of life, in every state in the Union, are represented in the ownership of the Bell Telephone System. People from every class of telephone users, members of every trade, profession and business, as well as thousands of trust funds, are partners in this greatest investment democracy which is made up of the more than 175,000 stockholders of the American Telephone and Telegraph Company.

If this great body of people clasped hands they would form a line more than 150 miles long. Marching by your door, it would take more than 48 hours of ceaseless tramping for the line to pass.

This democracy of Bell telephone owners is greater in number than the entire population of one of our states; and more than half of its owners are women.

There is one Bell telephone shareholder for every 34 telephone subscribers. No other great industry has so democratic a distribution of its shares; no other industry is so completely owned by the people it serves. In the truest sense, the Bell System is an organization "of the people, by the people, for the people."

It is, therefore, not surprising that the Bell System gives the best and cheapest telephone service to be found anywhere in the world.

"BELL SYSTEM"
AMERICAN TELEPHONE AND TELEGRAPH COMPANY
AND ASSOCIATED COMPANIES
One Policy, One System, Universal Service, and all directed toward Better Service

American Telephone & Telegraph Company advertisement, 1921. Taking the presidency of AT&T for a second time in 1907, Theodore Vail sought to provide customers "One Policy, One System, Universal Service." To achieve this goal, the company began buying up competitors. Seeking to be a government-sanctioned monopoly, AT&T touted itself as a benign "natural monopoly."

77

BRANDING

Manufacturers and the first generation of advertising professionals in the late nineteenth century created national brands with trademarked logos, characters, and package designs. Branding encouraged consumers to remember and seek out certain products over others and fostered an ongoing commitment to those products based on trust.

A successful brand, such as the one N. W. Ayer & Son created for the National Biscuit Company in 1899, reassured consumers about the standardized quality and consistency of the product. The campaign, including the famous Uneeda "Slicker Boy," encouraged consumers to buy crackers and biscuits made in factories and sealed in wax paper packages over the homemade variety or ones made locally and sold in bulk from barrels.

National Biscuit Company "Slicker Boy" ad, about 1900.

consumers. Earlier, companies, reacting to high fixed costs and the depression of 1893, had turned to forming large corporations, often acquiring competitors or merging together, in an attempt to bring order and an end to price wars. Between 1895 and 1904, about 1,800 American firms disappeared as a wave of mergers (mostly horizontal) swept through the manufacturing sector seeking to eliminate competition. The federal government passed the Sherman Antitrust Act in 1890 to take steps to stop monopolies and to insure competition. As President Theodore Roosevelt declared in 1902, "We draw the line at misconduct not against wealth."

LABOR

Workers and managers battled for control of the workplace under the new industrial system. As industry became big and relationships between workers and owners distant, workers created associations and unions to gain power and win demands. Strikes were one means of making employers pay attention. By organizing into unions, workers sought power to negotiate the pace of work, a less arbitrary hiring and firing system, safer workplaces, and a bigger share of the profits of productivity.

Organizing workers is the first and often most difficult step in creating a union, but how to go about the process is by no means straightforward. In the mid-nineteenth century, small local and regional groups of workers banded together. Later, labor organized around skilled craft or trade lines. Eventually, large national industrial unions strengthened by government protection became dominant. In the United States, efforts by socialists, communists, and anarchists to fundamentally change the capitalist system failed.

The growth of a shared laboring-class identity at the end of the nineteenth century helped bring workers together, but tensions relating to skill, race, gender, and nationality made the creation of large labor organizations challenging. Companies, such as those in the steel, mining, and railroads, used the courts, government, and coercion to discourage workers from forming lasting unions. By the 1930s, with the protection and standing offered by New Deal legislation and agencies, union organizers began to succeed.

The son of Irish immigrants, Terrence Powderly (1849–1924) became a railroad worker at age thirteen and a machinist at age seventeen. He joined the Knights of Labor in 1876 and was elected Grand Master in 1879. Over time he transformed the Knights from a fraternal society stressing the general dignity and value of work in a classless society to a national trade union open to men, women, and people of all races with the exception of Asians. In a motto of solidarity, the Knights of Labor argued "…an injury to one is the concern of all." While Powderly, like many other labor leaders of the time, believed in boycotts and arbitration, he opposed the use of strikes. He had, however, only marginal control over the actions of the membership.

The American economic depression that started in 1873 led to hard times, a rise in labor militancy, and a growth in the membership of the Knights of Labor. The Railroad Strike of 1877 was a key turning point. When the Baltimore and Ohio Railroad (B&O) cut wages for a second time in a year, workers walked out, and the first national strike took place. Violence broke out in the streets of Baltimore and elsewhere. Federal troops were brought in to quell the labor rebellion.

"The Baltimore and Ohio Railroad Strike," Baltimore, Maryland, *Frank Leslie's Illustrated Newspaper*, 1877. Most disagreements between management and labor were solved without strikes much less violent encounters. The first U.S. national strike took place in 1877. When violence broke out, federal troops were brought in to quell the labor rebellion.

above, left: **American Federation of Labor annual convention badge. 1896.** The American Federation of Labor (AFL) was an umbrella organization made up of workers from skilled trades. The union, led by Samuel Gompers, did not include unskilled laborers, women, or African Americans.

above, right: **IWW promotional sticker, about 1930.** Small recruiting posters like this one for the Industrial Workers of the World union were often put on boxcars to catch the attention of sympathetic itinerant workers riding the rails as they looked for jobs. The IWW welcomed all workers in the "one big union."

Powderly pushed with limited success for broad-based change: the eight-hour day, an end to child labor, equal pay for equal work, and political reforms like a graduated income tax. By 1886 the Knights of Labor, riding a wave of local victories, had 700,000 members. Growing labor violence, including the bomb explosion at the Haymarket Square rally in Chicago on May 4, 1886, triggered a nationwide wave of arrests and repression. By 1890 Knights of Labor membership had dropped to 100,000 and the union lost its position as the voice of labor.

Samuel Gompers (1850–1924) brought the American Federation of Labor (AFL) together mostly along craft lines, emphasizing local gains rather than political change. Unlike Powderly, he believed broad-based industrial unions were too diffuse and undisciplined to fight the repressive tactics of government and management and he pushed for more restrictive craft unions limited to skilled workers in a single trade. As Gompers said, "pure and simple unionism" (Foner and Garraty 1991). His goals, again unlike Powderly's, were for more immediate benefits– "more and more, here and now" (Rubenstein 1989). Facing strong court attacks and refusing to align with the Socialist party, Gompers pursed a nonpartisan policy. By 1920, the AFL had nearly 4 million members but lost ground throughout the decade.

Founded in 1905 by unionists, socialists, and anarchists, the Industrial Workers of the World (IWW) sought to abolish capitalism completely. Unlike the AFL, the Wobblies (as IWW members were called) organized by industry, paying no attention to skill, ethnicity, race, or gender. Never having a large membership, the Wobblies did have impact by making both government and industrialists worry about a class revolution. By the mid-1920s membership declined and the IWW ceased to be a factor in the American labor movement.

Unlike the IWW, the Congress of Industrial Organizations (CIO) was far more successful in organizing and winning demands. During the Great Depression, John L. Lewis organized the CIO in 1935 after the AFL chose to stay with a skilled craft union approach and not organize unskilled workers.

ADDIE CARD
1897–1993

In 1910 twelve-year-old Addie Card worked as a spinner in the North Pownal Manufacturing cotton mill in rural Vermont. She was born in Pownal, Vermont, in 1897; her mother died two years later and her father left town. Addie went to live with her grandmother.

Her image, however, became a symbol of child labor when Lewis Hine photographed her as part of the National Child Labor Committee's effort to persuade state governments to pass child labor laws. Card never knew she was the symbol for that cause. Living an unremarkable working-class life, Addie's photograph was famous throughout the nation.

Addie Card, North Pownal, Vermont, by Lewis Hine, 1910.

A. PHILIP RANDOLPH
1889–1979

Born and raised in segregated Florida, African American A. Philip Randolph moved to New York at age twenty-two. A master organizer, Randolph fought economic and racial inequality in very public ways. A pragmatic socialist, he did not promote the economic policies of black leaders like Marcus Garvey. He organized the Brotherhood of Sleeping Car Porters (a predominantly African American labor union), took up the cause of a segregated military, led the AFL-CIO as vice president, and fought for racial equality within labor unions. His organization of the 1963 March on Washington typified a life-long commitment to public demonstration for nonviolent change.

A. Philip Randolph, by John Bottega, 1963.

A large industrial union drawn from coal, steel, rubber, garment, and other industries, the CIO embraced unskilled as well as skilled workers.

Critical in labor obtaining power in America was the passage of the Wagner Act (the National Labor Relations Act of 1935). The economic crisis of the Great Depression eroded public support for business to regulate itself and gave the labor movement a chance. New Deal politicians enacted legislation regulating business and for the first time guaranteeing the rights of workers to organize, bargain, and take collective action including strikes if necessary. Where once federal troops had been used to break up strikes, now the government was committed to defending workers' rights to freely associate, form a union, and bargain collectively. Employers were prohibited from using unfair labor practices such as: interfering or coercing employees including freedom of association; discrimination in hiring based on membership in a labor union; retaliation against workers who file charges or testify; and refusing to bargain with representatives of duly elected unions. By the end of World War II, collective bargaining in America had taken hold, and a new era in the division of the spoils of productivity was about to begin.

Beer tray, about 1905. Unions were not always antagonistic. An ornamented beer tray from a Philadelphia Beer Drivers Union shows a worker and manager shaking hands and working together. German-speaking workers dominated many beer-related unions. The tray reads in German "Liberty, Equality, Fraternity."

PALACES OF CONSUMPTION

In the 1890s shopping had largely changed from a utilitarian activity to a social experience as middle-class and prosperous Americans embraced consumption as a means to exhibit class distinction. For aspiring members of the working class, shopping provided an opportunity to blur class lines. With one stylish dress a young working-class woman could imagine passing as a mannered person of means. The democracy of retail demanded that the merchant embrace both the carriage and shawl trades.

The urban department store, with its sumptuous architecture, set prices, and wide selections of goods and services first began to appear in the mid-1800s but did not take off until the 1880s. The stores encouraged middle-class and affluent consumers to spend the day enjoying the pleasures of shopping. Responding to the new inviting environments, a fusillade of advertising, and the advent of consumer credit, customers began to purchase more, especially durable goods. Merchants followed the lead of manufacturers and applied the economies of scale and division of labor to their stores. They invested enormous amounts of capital to build and stock the palaces of consumption and reaped significant profits for their marketing innovation.

One aspiring young entrepreneur, Marshall Field, moved to Chicago from Massachusetts in 1856 and began working his way up in the dry goods business. Partnering with different people, he was involved in several retail operations, eventually owning his own store, Marshall Field & Company, in 1881. Sensitive to gender concerns, Field realized that

Marshall Field & Company, Chicago, 1906. Around 1890, shopping in a department store became an experience, not a necessity, as predominantly upper- and middle-class American women began flocking to the stores. A relatively democratic institution, the free-entry and one-price policies in theory guaranteed the same reception and treatment for all.

ADVERTISING THE PALACE STORE

John Wanamaker, owner of Wanamaker's department store in Philadelphia, Pennsylvania, hired copywriter John E. Powers in 1880 to produce advertising copy for his store, then called the Grand Depot, to appear in local newspapers six days a week. Wanamaker, a devout Presbyterian, refused to advertise on Sunday. Powers's copy emphasized colloquial, direct, and candid language. One ad read: "We have a lot of rotten gossamers and things we want to get rid of" (Doyle 2013). The ads were so forthright that customers couldn't help but notice. The tactic worked; sales volume at the Wanamaker store doubled, making Powers' strategy an inspiration for others to follow. Powers explained the need to grab the attention of readers and the importance of sticking to the truth: "If the truth isn't tellable, fit it so it is" (Doyle 2013).

John Wanamaker Building, Philadelphia, Pa.

Wanamaker's department store, Philadelphia, Pennsylvania, about 1900.

the majority of his customers were women and that they should be catered to differently than men. Field is reputed to have coined the phrase "Give the lady what she wants." Between 1890 and 1940 Field, and the retail industry at large, shifted to an emphasis on service: one price (no haggling), accessible goods (not behind the counter), a tolerance for customer browsing, and easy returns. Department stores sought upper- and middle-class customers with impulse money to spend.

Working-class people could afford occasional department store purchases, but not luxury goods. More important, the working class lacked the time to browse the store as a leisure activity. Working-class people typically shopped in the neighborhoods where they lived and worked, where storekeepers often were kin or friends, where they could speak a familiar tongue, and where they might obtain credit. For working-class women, the department store window was more likely the destination than the store itself.

In the 1890s managers began to optimize the layout of department stores. High-profit impulse items, like cosmetics, were placed on the first floor, where female customers might be tempted while on their way to other departments. Managers also began changing the design of department store fixtures. Low display cases replaced high shelving, allowing customers to take in the whole store and browse the goods. While not focusing primarily on the working-class customers, Marshall Field & Co. created a bargain basement around 1905 to attract less affluent consumers. The bargain basement was also a place to sell overstocked goods. A carryover retailing practice of the earlier Merchant Era was the retailer's practice to hold stock indefinitely, until it sold. Slow-moving goods could grow dusty on the shelves for more than a decade. The bargain basement and sales did away with this approach. Over all, department store rules instructed sales clerks to be "polite and attentive to rich and poor alike."

In order to sell more, merchants also began to change credit terms. In the 1920s consumer debt accelerated as usury laws were changed and personal finance companies, making small loans to the aspiring working-class or low-level white-collar workers, became legal. Between 1916 and 1929 the number of small loans increased by over 3,000 percent. American advice literature of the late 1700s and early 1800s had warned about the evils of debt, with Benjamin Franklin writing, "Maintain your independency. Be frugal and free" (Calder 1999). It was Gilded Age authors (1870s to 1900) like Mark Twain, though, who began worrying that American society was changing from thrifty and debt averse to credit dependent.

The belief that Americans used to buy everything cash-and-carry but now use credit is largely a mistake, what historian Lendol Calder refers to as "the myth of lost economic virtue" (Calder 1999). Credit in retail sales had existed since the eighteenth century, just in different forms. In an agrarian society, debt was a means of evening out the cycles

Marshall Field & Company, Chicago, 1910. Window shopping was a popular pastime. The displays provided an opportunity for the carriage trade to be enticed, chat with their friends, and glance at their own reflections. The working class could see the latest styles without the intimidation of going into the store.

Cash register from Marshall Field & Company, Chicago, 1914. Department stores were often slow to adopt cash registers as supervising the operators for honesty and efficiency was difficult. Many stores preferred pneumatic tubes and centralized handling of money.

J.L. Hudson charge coin, 1919. Before plastic universal charge cards became popular metal charge coins served a similar function. Department stores like Detroit-based J.L. Hudson offered its better customers store credit accounts with special numbered plates. In most cases the account was to be paid monthly in full.

of income (many farmers sold their crops only once a year). In the new wage-based society, most workers got paid weekly or bi-weekly; debt (the result of credit) became a form of savings if the product purchased saved the consumer money. A car saved trolley fare, a washing machine made it easier to let go of the maid. In the 1910s and 1920s, mass finance became a pillar of a consumer society along with mass production and mass marketing.

As social taboos about personal debt slipped away, the use of consumer credit in the teens and twenties expanded, promoting a credit revolution. Many department stores issued store credit (payable at the end of the month) to favored customers. Credit purchases of durable goods (items that last for over three years) increased. In the 1920s, over 70 percent of new cars, 65 percent of used cars, 70 percent of furniture, 75 percent of radios, 90 percent of pianos, 80 percent of phonographs, 25 percent of jewelry, and 80 percent of household appliances were purchased with credit. Buying was made easier by the wages of industrialization: between 1860 and 1890s real wages (and disposable income) increased by 50 percent. Consumption based on credit forced people to work. Installment plans required specific payments (discipline of credit), which drove budgeting and routinized work.

FLORENCE KELLEY
1859–1932

Florence Kelley's father toured her through factories when she was a young child, and what she saw made a lasting impression, sparking a lifelong crusade against unfair working conditions. She studied law at the University of Zurich and was active in socialist causes. In 1891 she joined Jane Addams and other reformers at the Chicago settlement Hull House. As the first General Secretary of the National Consumers League in 1899, she fought for the eight-hour day, the minimum wage, and enactment of child labor laws. Her White Label program sought to inform consumers of products made under fair working conditions but was largely unsuccessful. Consumers' demand for labeled goods was limited and few manufacturers participated.

United Garment Workers of America advertisement, 1902. When the White Label program did not succeed, the National Consumers League switched their support to the Union Label.

HATTIE CARNEGIE
1886–1956

Seeking to distance herself from her ethnic origins, Henrietta Köningeiser renamed herself Hattie Carnegie after a different immigrant, Andrew Carnegie, the richest man in America. She opened a fashionable New York City hat shop by that name in 1909. Soon thereafter she added dresses that she herself designed, though she had no formal training and could neither draw nor sew. As a respected and successful tastemaker, she reinterpreted Parisian dress for the American consumer and was one of the first to introduce high-end ready-to-wear women's clothing.

Hattie Carnegie, about 1950.

Department stores were not only sites of consumption; they were also important as places of work. The stores rivaled large industrial factories, employing armies of workers in front of the counters selling products, and a large laboring class behind the scenes making the bureaucracy of the mammoth store function. Marshall Field & Co. was one of the first merchants to hire women as salesclerks. By 1898 R.H. Macy & Co. department store in New York had 3,000 total employees. In 1900 Jordan Marsh & Co. in Boston had between 3,000 and 5,000 employees and in 1904 Marshal Field & Co. in Chicago had about 9,000 workers. The department store helped change national employment patterns by offering jobs to a growing class of single female middle-class workers who then became consumers themselves.

TRANSFORMING AGRICULTURE

The first business of the nation was agriculture and American identity has long been tied to the concept of the yeoman farmer. In 1781 over 80 percent of Americans were farmers. President Thomas Jefferson provided insight on American attitudes toward agriculture saying, "Cultivators of the earth are the most virtuous and independent citizens." Despite nineteenth-century industrialization and the movement of many people to cities, rural life remained important and dreams of owning land and farming were still strong. As late as 1920, half of the U.S. population still lived in the country.

Homesteaders like the Sylvester Rawding family moved west into the arid prairie lands of Custer County, Nebraska, in the 1880s, pushing out ranchers who had earlier pushed out the native Indians. With the possibility of free or inexpensive land and little in the way of equipment investment, everyday Americans lived out the American dream of independence—and farm ownership.

In the early 1900s American farming began a major transformation, moving from extensive to intensive practice. The traditional approach had

Pioneer home, Custer County, Nebraska, photograph by Solomon Butcher, 1886. Sylvester Rawding with his wife Emma, sons Philip, Harry, and Willie, and daughter Bessie pose proudly outside their simple prairie sod home. Rawding was born in England in 1828 and at the time of the Civil War he lived in Illinois.

been to plow up more land in order to increase production, but throughout the 1800s yield per acre only inched up. More farmers on more land meant more food, but making those farms more efficient was yet to come.

A continuing series of innovations in machinery, seeds, and chemicals increased efficiency and rocked the outlook of rural America. As farmers moved west, plows that were suitable for the tough prairie soils of the Midwest became necessary. In 1848,

Windmill patent model, 1880. Windmills, like Henry Beville's "Iron Duke," provided farmers and ranchers with a power source to pump water from underground. Between 1870 and 1900, about 230 million acres were put into agricultural production, much of it in the arid Great Plains.

John Deere (1804–1886) began producing plows in Moline, Illinois. He manufactured a steel plow, which he had first developed in 1838. In 1868, James Oliver (1823–1908) opened the Oliver Chilled Plow Works in South Bend, Indiana, selling a patented case-hardened cast-iron plow.

Horse-drawn equipment, like plows and reapers, eliminated certain production bottlenecks and reduced labor needs but did not increase yield per acre. Of course the purchase of equipment required money and some farmers began taking on debt. The need to borrow money for agricultural equipment helped soften American social taboos about debt, opening the door for the consumer credit revolution of the 1920s.

While new equipment made farmers slightly more productive and the work less tedious, there

Fordson tractor, about 1924. As lightweight tractors dropped in price in the 1910s, farmers began to move away from horse-drawn equipment. At its peak in 1923 Ford produced 101,938 Fordsons. Tractor farming increased the capital requirements to start a farm, gradually making it impossible for the poor to elevate themselves.

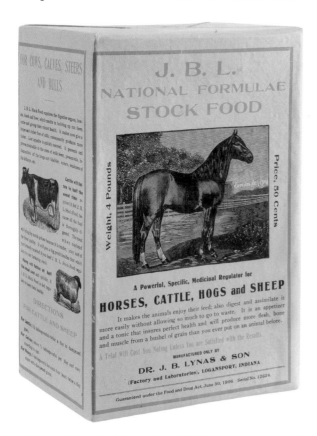

Dr. J. B. Lynas & Son Stock Food, about 1910. Horses and mules provided power for agricultural equipment but caring for and feeding of the animals was expensive. The J. B. Lynas firm was located in Logansport, Indiana, and in 1913 employed about 1,400 people. The company sold beauty products and patent medicines.

were drawbacks to animal power. Cost was one of them: farmers had to buy horses and mules, dedicate land to feed the animals (six acres to grow the feed for a single horse), and pay for veterinary care. While labor hours dropped, per acre productivity on American farms rose very little until the 1940s.

In the late nineteenth century, government support and education began to improve farming techniques, crops, and productivity. Founded in 1862, the United States Department of Agriculture (USDA) sent seed explorers around the world in search of new or better cultivars. Samples were distributed to farmers for trial. In 1897 the USDA distributed over 20 million seed packages. Government seed explorers Palemon Howard Dorsett and William Joseph Morse were dispatched to Asia from 1929 to 1932 to study soybeans, and they sent home about 9,000 germplasm samples. The USDA, in turn, introduced soybeans to American farmers—not for crop cultivation so much but primarily for soil enrichment. After World War II the usefulness of the bean itself was developed.

The federal government also helped improve agriculture by funding agriculture colleges. The Morrill Acts of 1862 and 1890 gave federal land to the states so that the sale could be used to fund agricultural, "land grant" colleges. Increasingly, government encouraged farmers to become modern business managers and to take greater risks in order to produce more crops. The schools trained agronomists and did much research into hybridization of both plants and animals. By the 1920s, private seed hybridizers began to replace the federal government as the source of new seeds.

Threshing wheat, early 1900s. Even with mechanization, harvest time was an intense period of work for farmers. Here, a group of farmers and hired workers use threshing machines connected by large leather belts to steam-powered traction engines to thresh the year's wheat harvest.

TEI SHIDA SAITO
1896–1989

Tei Shida, born in Fukushima, Japan, was a picture bride. Her parents had arranged her marriage to Japanese Hawaiian immigrant Saito without her knowledge. Leaving Japan with him in 1913, still a teenager, Tei Shida, like other Japanese picture brides, found her new life isolating and lonely. She joined thousands of other Japanese immigrants who provided much of the labor for Hawaiian agriculture. Her Japanese upbringing had not prepared her for the laborious work helping manage a pineapple plantation.

Tei Shida at the time of her high-school graduation, Fukushima, Japan, 1912.

The old agricultural way of life, built on animal-powered equipment, saving seeds, and traditional practice, was fading away by the early 1900s. The use of large steam- and oil-burning traction engines powering threshers did not at first bring about widespread change, however. The advent of inexpensive lightweight tractors was different. During World War I, the large demand for agricultural products coupled with a scarce labor supply resulted in high crop prices. Lightweight tractors were seen as one way to meet demand with fewer workers. There were 15 tractor manufacturers in 1910, 61 in 1915, and a staggering 186 in 1921.

Raised on a farm, Henry Ford entered the tractor business in 1917. "I have followed many a weary mile behind a plough and I know all the drudgery of it. What a waste it is for a human being to spend hours and days behind a slowly moving team of horses when in the same time a tractor could do six times as much work!" (Ford and Crowther 1922). The notoriety of Ford as the genius developer of the Model T and a low price for the Fordson tractor made possible by mass production helped the Fordson gain market domination (nearly 75 percent market share in 1923). The initial price for the Fordson tractor in 1918 was $750, but with production improvements and a price war with International Harvester the price dropped to $350 by 1922. Yet Ford's unresponsiveness to customer feedback and unwillingness to update the Fordson tractor led to its failure and departure from the U.S. market in 1928.

BOOTLEGGING

Prohibition in the United States began with the enactment of the Eighteenth Amendment to the Constitution (the Volstead Act) in 1920. While the banning of most alcohol from 1920 to 1933 did not stop widespread drinking by Americans, it did create a large market for illegal alcohol. Passed as an effort of common good, Prohibition had the unintended consequence of institutionalizing violent organized crime. From counterfeiters and bank robbers to cattle rustlers and con artists, Americans had broken laws in the past. But the widespread public desire for alcohol during Prohibition enabled the business of crime to flourish as never before. Illegal sources of production and distribution, bootleggers, quickly sprang up and mob bosses like Alphonsus "Al" Capone (1899–1947), Charles "Lucky" Luciano (1897–1962), and Arthur "Dutch Schultz" Flegenheimer (1901–1935) turned to violence to control competition. Crime became big business and expanded from bootlegging to gambling, prostitution, extortion, and other unlawful acts.

One of the most notorious mob bosses, Al Capone, ruled Chicago with an iron fist. Because of his charitable donations and the unpopularity of Prohibition, Capone was popular with many people. After the violent and widely reported St. Valentine's Day Massacre of rivals in Chicago in 1929, Capone's popularity tumbled.

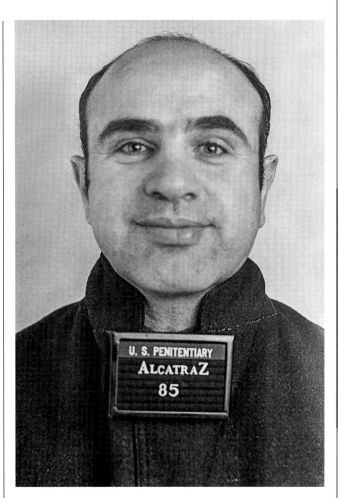

Mug shot, 1931. Born in Brooklyn, New York, in 1889, Alphonsus Capone was active in New York gangs. He moved to Chicago in 1919 and ran a large bootlegging, gambling, and prostitution operation. A good businessman, Capone ran an efficient albeit violent organization. He was convicted of tax evasion in 1931.

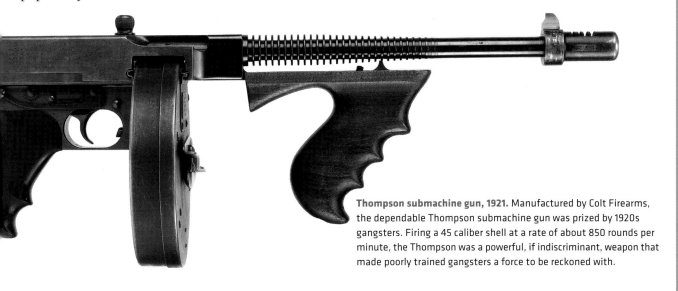

Thompson submachine gun, 1921. Manufactured by Colt Firearms, the dependable Thompson submachine gun was prized by 1920s gangsters. Firing a 45 caliber shell at a rate of about 850 rounds per minute, the Thompson was a powerful, if indiscriminant, weapon that made poorly trained gangsters a force to be reckoned with.

CHARLES PONZI
1882–1949

Charles Ponzi, after whom shady pyramid schemes are named, promised prospective investors fantastic investment growth but based his fund primarily on enrolling other people into the scheme. He built his deceit on complicated fluctuating foreign currency rates for international postage coupons, which was legal. In the early 1920s, he promised investors huge returns for their monies, but his business never brought in substantial assets; rather it used new investors' money to pay off old investors. When finally arrested in 1920, his prosecution left 30,000 investors nearly $20 million in the hole. Undeterred, after several years in prison, Ponzi sold investors swampland in Florida—another scam for which he was indicted and eventually charged.

Charles Ponzi working at a desk, 1920.

above and opposite: **Dougherty's Whiskey for medicinal use and medicinal whiskey prescription.** While the sale of alcohol was banned during Prohibition, there were some loopholes and very little public support for the banning of alcohol. Many people got doctors' prescriptions for medicinal liquor.

opposite, bottom: **Cocktail set, 1924.** Prohibition may have shut down some saloons, but many people continued to enjoy cocktails and other forms of alcohol at home. This stylish cocktail set was used during Prohibition by the Reader family of West Hatton, Maryland.

THE GREAT DEPRESSION

Up and down business cycles are typical in any economy. The United States had significant depressions, panics, and economic crises in 1837, 1857, 1873, 1893, 1907, and 1914. But the depth of the problems during the Great Depression gave American capitalism one of its greatest tests ever. Worldwide events, plunging agricultural markets, widespread unemployment, and the financial chaos of the stock market crash and bank failures made Americans question the wisdom of allowing business to self-regulate.

The Great Depression was brought on by many causes with lax regulation and foolish optimism being significant. In the 1920s firms overinvested in capacity, consumers turned to credit in large numbers, farmers thought demand would continue forever, banks departed from traditional conservative practice, and risky stock speculation spread to middle-class citizens incapable of absorbing big losses.

Economic crisis hit American farmers in the 1920s. Following the end of World War I, world demand for food and fiber dropped precipitously. Widespread drought and debt (that farmers had piled up thinking the good times would not end) made things worse. In a perfect storm of problems, the shift in agriculture to motorized mechanization in farming reduced labor needs, throwing many people off the land.

The American financial system began to unravel in late 1929 as an extended stock bubble (where prices rise without regard to value) burst and many leveraged speculators couldn't meet their debts, accelerating the stock market crash. In addition, a lack of financial regulation had left banks susceptible to market swings. Some banks did not

*opposite: **Migrant Mother**, by Dorothea Lange, 1936.* This photograph of Florence Owens Thompson at Nipomo, California, became an iconic image of the Great Depression. While many farmers were displaced by drought, many others were thrown off the land by mechanization. Both groups joined unemployed city workers migrating to California hoping for jobs.

COMMERCIAL RADIO

As radio programming developed in the 1920s and 1930s in the United States, broadcasters created a system funded by advertising. Commercial radio came into the home and built intimate relationships between sellers and consumers, filling living rooms and kitchens with the friendly voices of radio pitchmen and women. Relationships were formed, and on a daily basis listeners could tune in to trusted friends advising them about what brands to buy. As the boom years of the 1920s gave way to the Great Depression, radio personalities shared the hardships of listeners, and advertising turned to the "hard sell" that drove home the economical features of products.

Philco radio, 1930s.

PORTIA HOWE SPERRY
1890–1967

Losing a job and with little hope of finding a new one was a reality that many people shared in the 1930s. When Ralph Sperry was laid off as a piano designer, the family experienced failure. Undaunted, Portia Sperry got creative and made opportunities for herself and others. The family moved to rural Nashville, Indiana, and Portia found a part-time job selling handmade objects from the area. Hoping to generate income for the family, the entrepreneurial Portia Sperry designed a doll based on native crafts. She assembled a network of home workers to manufacture it, and with entrepreneurial gusto convinced Marshall Field & Co. to sell the doll. Her story of cooperative self-help resonated with the nation and was a marketing success.

Portia Howe Sperry holding one of her dolls, 1934.

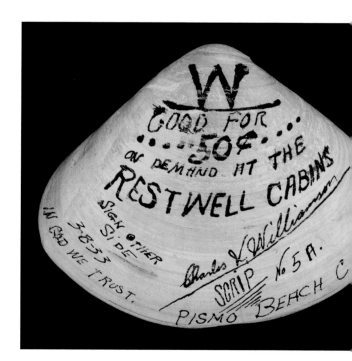

Emergency money, 1933. The Great Depression put the American financial system into turmoil. Without sufficient legal tender in circulation, some companies had to create their own currency to make change. Leiter's Pharmacy in Pismo Beach, California, issued specially painted clamshells as change.

have sufficient cash reserves to meet depositor demands, and in December 1930 there were a number of bank failures. The subsequent dramatic drop in the money supply and a rise in credit rationing left businesses without enough money to operate, feeding unemployment and making a bad situation worse.

Unemployment is common in an economic downturn, but the extent and length of the unemployment crisis during the Great Depression were unique in American history. At its height, in 1933, nearly 25 percent of the labor force was jobless, and unemployment would stay above 15 percent throughout the 1930s. With little or no income and the fear of losing their jobs, workers tightened up spending, reducing demand. The downward cycle of the Depression deepened.

***Oct 29 Dies Irae*, by James N. Rosenberg, 1929.** Artist James Rosenberg captured the panic that swept the streets of New York City on Black Thursday, October 29, 1929, on this lithograph, as the stock market crashed. In fact, while the artwork is emotionally correct, the market tumbled over many days.

THE NEW DEAL

With the election of Franklin Delano Roosevelt as president in 1932, the federal government responded to the perceived failure of capitalism during the Great Depression with a wide array of government initiatives known as the New Deal. Without a single ideology or plan, the Roosevelt administration initiated a series of programs establishing a larger government, one more active in business affairs and willing to regulate aspects of free enterprise in favor of the common good.

The programs fell largely into two categories, recovery and reform, and were enacted in three phases—first 100 days, 1936 reelection, and second-term recession. In 1933 the government acted quickly to stop the economic panic, in the second two phases it sought more general reforms. President Roosevelt created an "alphabet soup" of regulatory agencies to prop up the banking system, reform the stock market, aid the unemployed, and induce industrial and agricultural recovery. Some actions were spectacular successes, some needed refinement, and some were failures. Public and business opinions about the New Deal continue to vary; some see it as savior, others as destructor.

The 1920s were a turbulent time for the financial markets. Six hundred banks failed each year between 1921 and 1929, a rate ten times higher than that of the 1910s. Matters got much worse in the 1930s, with a record-breaking 4,000 bank failures in 1933. The Federal Deposit Insurance Corporation (FDIC) was created in 1933 to rebuild public trust in banks by guaranteeing deposits and tightly regulating banks. It was also a time of significant stock speculation and manipulation. The Securities and Exchange Commission (SEC) was created in 1934 to regulate the nation's securities markets. The markets got better very slowly, and in 1938 the Federal National Mortgage Association (Fannie Mae) was created to pump federal money into the home mortgage market, starting the tradition of government action to expand home ownership and the availability of affordable housing. In President Franklin Roosevelt's 1933 inaugural address he proclaimed, "There must be a strict supervision of all banking and credits and investments; there must be an end to speculation with other people's money, and there must be provision for an adequate but sound currency" (George Mason University).

Business viability and employment were other major problems. The National Recovery Administration (NRA) was established in 1933 to regulate production, prices, and wages. At first supported by industry and labor, both groups soon became disenchanted with the NRA. In 1935 it was ruled unconstitutional by the Supreme Court.

Seeking labor stability after over 500,000 workers, from a wide variety of industries, went on strike in 1934, Congress passed the Wagner Act,

NRA Blue Eagle, about 1930s. The National Recovery Administration (NRA), one of the centerpieces of the New Deal, was established in 1933 to regulate production, prices, and wages. The hope had been to create a government-managed economy and end destructive competition. The Supreme Court declared the NRA unconstitutional in 1935.

Political cartoon, by Clifford Berryman, *Washington, D.C. Star*, **1938.** President Franklin Delano Roosevelt attempted to revive the U.S. economy by stimulating the economy and creating recovery programs and regulatory agencies to bring about permanent social change.

establishing the National Labor Relations Board (NLRB) in 1935. After two years of ambiguity the Supreme Court ruled employees could organize and unions were not an antitrust violation.

One of the few New Deal organizations that set out to directly lower unemployment, the Works Progress Administration (WPA), formed in 1935, hired people to work on public work projects.

Agriculture and rural life was the focus of much New Deal activity. The Agricultural Adjustment Administration (AAA) attempted to raise farm income and improve soil conservation through subsidies, purchase programs, and education. Congress sought to regulate supply and demand for seven basic crops.

Perhaps one of the most fundamental and long-lasting New Deal acts was the creation of the Social Security Administration (SSA), passed in 1935. The new agency provided a safety net of retirement, disability, and survivors' benefits. Social

Union buttons, 1930s. The passage of the National Labor Relations Act (1935) for the first time ensured labor's right to organize. It guaranteed the rights of workers to organize, bargain, and take collective action. It also meant that employers could no longer legally coerce or intimidate their employees.

Security shifted some of the responsibility for the poor, elderly, and infirm from local communities to the federal government. It began an important entitlement and an equally important long-term fiscal liability.

Social Security along with other New Deal legislation and programs provided for a more regulated and predictable economy. The failure of business to effectively self-regulate prior to the Great Depression challenged the public and politician's confidence in laissez-faire economics. The New Deal programs and ensuing years of control demonstrated a shift in the public perception of the appropriate balance between state decision-making and individual entrepreneurial choice.

The Corporate Era was a time of dramatic change for government, business, workers, and consumers. In the early years of this period, the federal government won the Civil War and preserved the union. In the process, it became stronger and more involved in the nation's business and finance. It established a stable national paper currency, the "greenback," opened more land for settlement with the Homestead Act, and expanded education with the Morrill College Land-Grant Act. In a major new initiative, it fostered completion of a transcontinental railroad by passing major Railroad Acts in 1862 and 1864. Following the war, westward expansion boomed.

Between 1870 and 1900, American agriculture doubled. Farms grew in numbers from 2.6 million to 5.7 million, in acres under till from 407 million to 841, and the value of crops rose from $9.4 billion to $20.4 billion. But by 1920, this growth had peaked. Where 41 percent of the workforce was employed in agriculture in 1900, only 16 percent were on farms by 1945. Workers were flocking to cities. By 1920 the American urban population was 50 percent of the nation. The dominant change of the era was the rise of factories and the

O.W. Olsen farm, South Dakota, 1936. Drought and poor farming practices turned much of the High Plains into a dust bowl, adding to the severity of the agricultural depression. New Deal programs sought to educate farmers on sustainable agricultural practice and support soil conservation.

reorganization of work. Railroad construction, shipbuilding, textile manufacture, production of consumer goods, and a host of other enterprises emerged across the nation. Inside these institutions, an even broader transformation of labor stemmed from three factors: substitution of semi-skilled labor for more expensive skilled labor, closer supervision of work, and intensification of pace. Self-directed craftsmen had sold complete products; workers under the new industrial system sold their time. The efficiencies of production allowed prices to tumble and the standard of living to increase. The Corporate Era was dominated by a new national identity focusing on materialism.

The success of big business raised concerns because wealth and power were concentrated on an unprecedented scale. By the turn of the century, muckrakers were warning of the trusts and monopolies and the slumbering government began to respond with timid legislation. Unfazed business continued to become large, impersonal, and increasingly run not by owners, but by a new class of professional managers. Capitalism seemed poised to overpower democracy. In the minds of some, the Great Depression of the 1930s proved how devastating the result could be. Government stepped in with strong regulation and social programs. Another war, World War II, also helped the nation regain its feet. Emerging from the recovery would be a new era in which the powers of business, labor, and government were redistributed, and consumption became ever more central to economic growth.

How the American Corporation Changed Everything

ADAM DAVIDSON

Adam Davidson is an award-winning journalist who, after reporting from Baghdad as a foreign correspondent for public radio, went on to focus on financial issues. A radio program he co-wrote and co-produced on the housing crisis during the Great Recession, *The Giant Pool of Money*, was named one of the top ten works of journalism of the decade by the Arthur L. Carter Journalism Institute at New York University. Davidson is co-founder of NPR's *Planet Money*. He also writes the monthly "On Money" column for the *New York Times Magazine*, where he explores how money and economics influence all of our lives in big and small ways.

Our lives today are an invention of the Corporate Era. So much of what we have come to see as simple and obvious actually represents the radical outcomes of a series of experiments conducted in the late 1800s and early 1900s by businessmen (they were, almost entirely, men) confronted with challenges nobody had faced before. The corporations that came into being after the Civil War were larger institutions than had ever existed in human history, coordinating activity—in real time—across far vaster space than had ever been possible. The Catholic Church, a few of history's largest armies, and the construction of massive monuments in Egypt, Mesoamerica, and China may have come close in raw numbers of workers, but each had developed a decentralized system of autonomous units reporting only occasionally to headquarters (President Lincoln's increasingly frustrated letters to his generals reveal just how little command the commander in chief wielded). Then, in the United States, almost as soon as the telegraph and the transcontinental railroad appeared, a new type of organization arose. Corporations needed to direct the activity of tens of thousands of workers spread all over the United States and, often, the world. Businessmen scoured the record, studying military histories, the Bible, church records, anything they could to find some sort of model. They found nothing they could apply wholesale and invented something new: the modern American corporation.

To demonstrate just how radical the innovations of that era were, consider something as basic as a job. Before the Corporate Era, almost nobody in history had a job, at least not in the modern sense of that word. People worked, of course. But they didn't have long-term commitments with an employer to be paid a set salary in exchange for a proscribed number of hours of labor. In a nation that had been mostly farmers and home-based merchants and craftspeople, the corporate job meant that work became a discrete activity, removed from the home and the family. This would have enormous impact on the nature of our personal lives. Family size fell. Farmers and cottage industry workers had wanted as many children as possible to provide more labor and to care for aged parents. But in an economy built around corporate jobs, children became a net cost, since they often left home precisely when they started earning their own wages. And those children increasingly had something few people ever had before: a choice of what to do for a living. Most people in America before the Corporate Era, like most people in human history, did whatever it was their parents did. But these large corporations and the cities they built around themselves offered young people a type of freedom that had previously been largely reserved for the younger children of the wealthy. No wonder the Corporate Era brought with it another new term: adolescence. America's leading psychologist of the age, G. Stanley Hall, identified a previously unknown phase of life, a period in which young people needed to experiment in order to properly make decisions in a world suddenly filled with new options.

Alongside the modern job, the Corporate Era also brought with it several close corollaries, like careers and professions and professional degrees and licenses. (Not long before, people didn't need anyone's permission to declare themselves doctors or lawyers or accountants.) These inspired the growth of education. In 1900, fewer than 10 percent of Americans graduated high school, and many children worked full time by the age of ten. Soon, a high-school degree would be the norm and child labor illegal.

As corporations grew in size and influence, government responded in kind. At the beginning of the Corporate Era, federal and state governments barely touched the lives of most Americans and had almost nothing to do with business. But in response to the rise of powerful firms and their sometimes despised leaders, politicians—most notably Teddy Roosevelt—reimagined the role of elected officials as a bulwark against this new powerful force in the country. Regulations grew, then came corporate and individual taxes, and the now-familiar and unending cycle of conflict and cooperation among civic advocates, politicians, and corporate lobbyists.

The Corporate Era represents a defining break between an old way of living and the new one we now know. Before it, even as late as 1850, the daily rhythms of life, the scope of one's attention, and the nature of family and personal relationships still had much in common with the medieval world. Even in the advanced factory towns of eastern Massachusetts, where the most cutting-edge industry lay, life was conducted at a slow amble. Machinery was powered by water; industrial goods were transported by mule cart or slow-moving steam train to port towns. Business was conducted by hand-written letters that would take weeks to reach trading partners. Factory workers disappeared during busy harvests or on an especially lovely spring afternoon. The weather, indeed, was a far more potent influence on people's successes and failures than far-away government or business decision-makers. Life was smaller, slower, more local.

In 1931, when he was twenty years old, Ronald Coase spent a year traveling the United States, visiting as many examples as he could of this brand-new and very American invention: the large, manager-led corporation. He was a not especially promising British undergraduate seeking the answer to a deceptively simple question: Why do these large companies exist? It was a confounding puzzle. Not long before, America seemed the perfect model of the ideal capitalist system as elegantly described by Adam Smith. It was an economy of small, nimble businesses constantly adjusting prices to perfectly match supply and demand, producing a socially optimal outcome. There was no central manager. A capitalist nation would become richer by freeing each person to pursue his or her own self-interest. Yet, suddenly, America, the very beacon of capitalism at the dawn of the American century, was dominated by these massive, lumbering, and centrally controlled institutions. These large corporations had entire floors filled with accountants and draftspeople and telephone operators, who spent their days doing whatever it was some manager in a central office told them to do. (If they weren't spending their days avoiding doing whatever it was their manager had assigned them.) How could corporations be so rich? How could they afford to pay workers more than they could make anywhere else? What had become of Thomas Jefferson's nation of independent, yeoman farmers? John Adams' nation of self-reliant citizens? After fighting for freedom, why had they handed over their sovereignty to become drones working indoors, controlled by powerful bosses they rarely ever saw?

On that trip, Coase developed the Theory of the Firm, a defining insight that inspired much later work, including the masterpiece of American business history by Alfred D. Chandler, Jr., *The Visible Hand: The Managerial Revolution in American Business*. Together these works showed that the modern corporation is THE great American invention. More than the telegraph or the phonograph or electric lights or the Model T, it's the corporation itself that has made America so rich. Of course, the Corporate Era did more than just bring wealth. It changed everything.

Can Business Serve Society?

BILL FORD

William Clay "Bill" Ford Jr. is in the singular position of being a descendant of the iconic American inventor and visionary Henry Ford and running the company his great-grandfather founded. Bill Ford joined the Ford Motor Company soon after college and held a variety of domestic and international assignments in manufacturing, sales, marketing, product development, and finance before being named chairman in 1999. He helped steer the company through the gut-wrenching Great Recession, when Ford famously survived while turning down government aid. Today, as executive chairman, he continues to push the company to be a leader in sustainability and technological innovation.

On a summer night in 1896, after years of experimenting, my great-grandfather Henry Ford finished building his first car. In that moment of triumph, he discovered the car was too big to make it through the door of his workshop. Without hesitation, he knocked down a brick wall and set out on a test drive through the streets of Detroit.

The modern age had begun.

Growing up in the Ford family, we regularly talked about the business around the dinner table and learned about my great-grandfather's achievements. I shared his love of cars and deep appreciation for the natural world and embraced his conviction that progress was driven by innovation. I also came to believe that his greatest contribution was one of the most overlooked: his belief that business must serve society.

At a time when business models were often based on exploitation, Henry Ford believed that a company should exist not simply to make a profit, but to make people's lives better. His vision was to grow by serving people, including customers, employees, and communities.

This began early on. In the decade that followed the assembly of his first car, Henry Ford would use a unique combination of vision and problem solving to create a series of progressively better designed, more affordable cars. In 1908, he introduced his automotive masterpiece—the Model T.

The Model T was affordable and reliable. It changed the way we live, work, and play, providing mobility and prosperity to people around the world. The thinking behind the car was as simple as the Model T itself: Henry Ford wanted to improve people's lives by building cars that were attainable for the average family. A later advertisement captured his vision in the clearest of terms: "Opening the Highways to All Mankind."

Innovation comes in waves, and the Model T was no different. Within a few years, Henry Ford introduced the moving assembly line and $5-a-day wage, ideas that worked to define the industrial age. For the first time, workers shared in productivity gains and could afford to buy the products they made. This remarkable combination of process and product put the world on wheels and improved the standard of living for millions of people around the world.

Between the start of Model T production in 1908 and its conclusion in 1927, more than 15 million vehicles were built. The Model T was the first car to be sold on six continents. It became an integral part of everyday life, of popular culture, and even our language. By the early 1920s, half of all cars in the world were Model Ts. Modern life was well under way.

A little more than a century after Henry Ford drove that first car on the streets of Detroit, *Fortune* named him Businessman of the Century. In bestowing this honor, they wrote, "No fewer than three of the biggest management brainstorms of the century happened in Ford's head: the idea of a moving assembly line, the idea of paying workers not as little as possible but as much as was fair, and the idea of vertical integration that made Ford's River Rouge plant the chief wonder of the industrial world. The oil industry, the highway-construction industry, nearly universal homeownership—all these things, from Big Oil to Big Macs, can trace their parentage to the Model T Ford. The American Dream itself is inextricably linked to the automobile."

Henry Ford was an iconoclast though, particularly when it came to his understanding of business. He fought with his partners to build a car that would enhance the lives of everyday people, not just the affluent. He insisted on investing profits into building a better, less expensive product and sharing those profits with employees. Henry Ford expressed this belief by saying, "There is a most intimate connection between decency and good business. The only foundation of real business is service."

While this holistic philosophy was counter to conventional business wisdom of the day, it is why he was able to build a great enterprise that provided previously unimagined levels of mobility and prosperity in the twentieth century.

My faith in this philosophy has been tested over time. When I began working in the auto industry in 1979, most industrial businesses put profits ahead of environmental concerns. I believed, and still do, that environmental sustainability is the most important long-term challenge for businesses around the world. This belief grew in part from my family perspective. When your family's name and reputation is connected to your business, you do what is right for all of your stakeholders. This means going beyond current stakeholders to embrace the responsibility we have to future stakeholders including our children and grandchildren.

Early on, when I spoke out on this issue, my views were not widely accepted. Many people in the business world thought I was eccentric or naïve. They feared that associating with environmental groups would be harmful to a company's reputation. Many environmentalists were skeptical of my motivation and sincerity.

But instead of pulling away from the groups I belonged to, I engaged more fully with them, speaking at conferences and meeting to discuss ways to move forward. Gradually, both sides found common ground and we were able to make significant progress in designing and manufacturing vehicles that customers want in more sustainable ways.

Over time, the thinking in the corporate world changed as the impact of global environmental crises became more apparent and the business case for stewardship more widely understood. Environmentally responsible companies demonstrated that they could improve their reputations and customer loyalty, along with their bottom-line results.

If last century's challenge was to put the world on wheels, the challenge now is to keep the wheels rolling. With a growing global population and greater prosperity, the number of vehicles on the road could double by midcentury, exceeding 2 billion. At the same time, the majority of the world's population will live in cities. The world's infrastructure cannot support such a large volume of vehicles without creating massive congestion and serious consequences for our environment, health, economic progress, and quality of life.

No matter how clean and efficient vehicles become, we simply cannot depend on selling more of them as they function today. Automobiles must become smarter and more integrated into the overall transportation system. This requires a change in our view of not just the car itself, but the way the car interacts with our society. It also requires a rethinking of the traditional business model to include collaboration across industries and with local and national governments.

Cars of the future will be mobile communication platforms that talk to each other and to the world around them, making driving safer and more efficient and potentially extending the driving lives of many people. Vehicles will be integrated into the transportation ecosystem in ways that optimize the entire system, interacting with a city's infrastructure and multiple modes of transport. Early stages of this transformation already are under way, with wireless connectivity, infotainment systems, and limited functions for assisted driving and parking being developed.

One hundred years ago, Henry Ford redefined mobility for average people, and we have the opportunity to do the same now. In the twenty years following that first test drive, the world was dramatically changed by the automotive industry. Now, we are on the verge of a similar transformation. It will present new ways of ensuring that my great-grandfather's dream of opening the highways to all mankind lives on into the twenty-first century and beyond.

The most successful corporations of the future will follow a similar pattern, moving beyond short-term thinking and narrow self-interest to make the world a better place.

Brownie Wise demonstrating Tupperware in Hawaii, 1950s. Hawaii, soon to become the 50th U.S. state in 1959, served as a dramatic backdrop for a Tupperware demonstration and publicity photo. Americans associated the tropical location with "the good life," something that Wise hoped would motivate salespeople and consumers.

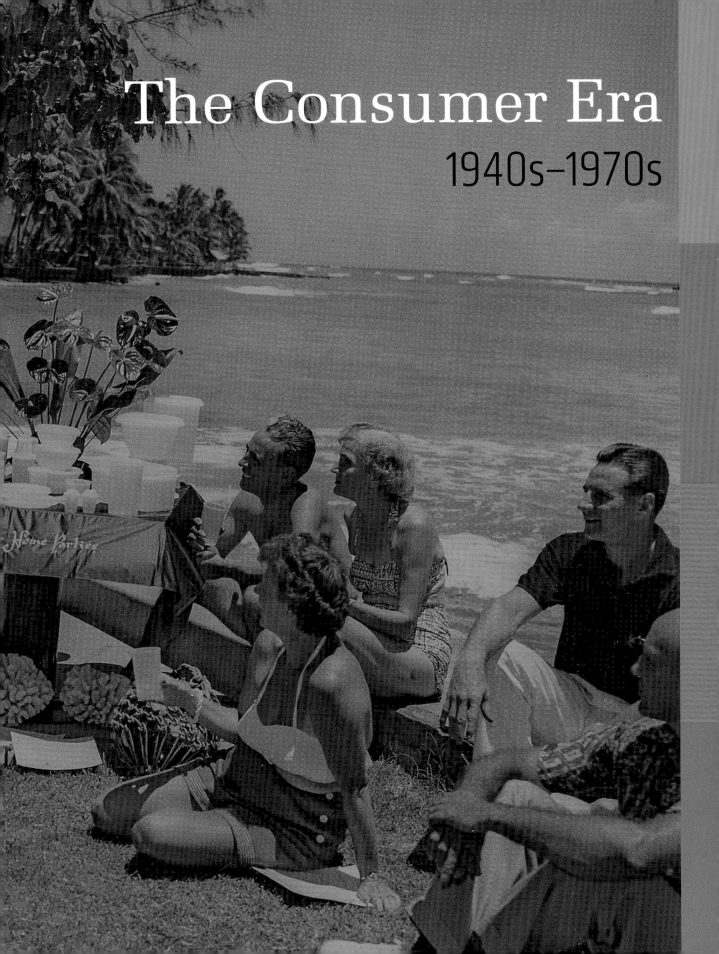

The Consumer Era

1940s–1970s

Hotpoint Combination Refrigerator-Freezer and product literature, 1959–1961. Refrigerators in a wide range of colors and styles symbolized a new era of American affluence. Sales literature turned the home appliance into a veritable cornucopia of fresh and frozen foods that underscored the productivity of American farms and the increasing food choices available to American consumers.

PROPONENTS OF CAPITALISM and communism met in an unlikely place in 1959. Vice President Richard Nixon and Soviet Premier Nikita Khrushchev squared off in front of a high-tech "Miracle Kitchen" and other material evidence of America's industrial and agricultural power in Moscow's Sokolniki Park. The U.S. Department of State opened the American National Exhibition, the first of a series of traveling exhibitions mounted by the United States in the Soviet Union to improve relations and display the fruits of American capitalism: convenient and abundant food, modern appliances, automobiles, an IBM computer, and color televisions galore. All of these innovations shared space under the golden dome of the exhibit pavilion, but it was the "Miracle Kitchen" that took center stage. With modern appliances that supposedly freed American women from time-consuming tasks as background,

Nixon and Khrushchev debated the role of government in facilitating prosperity. The debate would extend beyond the galleries of the exhibit hall and span several meetings between the two leaders as they attempted to win hearts and minds but also underscore their respective nation's commitment to a particular economic system—one that left business to private owners and the individual profit motive, with some state regulation; the other that placed production and business under the control of the state in order to distribute wealth more evenly.

The kitchen debates, however, were only one episode in a long series of discussions about capitalism and democracy, individual opportunity and the common good in the United States. In the postwar period of abundance, the benefits of capitalism seemed clear to many, if not all Americans, especially as those benefits took material form.

Richard Nixon and Nikita Khrushchev, American Exposition, Moscow, July 24, 1959. Nixon and Khrushchev argued the merits of capitalism versus communism in front of a model American kitchen. Nixon pressed Khrushchev on the shortcomings of the communist system and incited what became known as "the kitchen debates."

WILLIAM H. WHYTE
1917–1999

An editor at *Fortune* magazine and author of the provocative best-seller *The Organization Man* (1956), William Whyte argued that the organizational culture of big business in the postwar period threatened the entrepreneurial vigor and boldness of American businessmen, especially middle managers, who strived for conformity and stability rather than individualism. Whyte, one among a chorus of voices that worried about the deadening effects of bureaucracy, encouraged Americans to buck the organizational mindset to reinvigorate business.

William H. Whyte, mid-1950s.

Americans were no strangers to these messages; capitalism as the heart of American enterprise and the source of affluence threaded its way through advertising, film, and television. In reality, the Consumer Era encompassed a high point of productivity. World War II spurred innovation and diminished global competition, laying the foundation for American dominance in technological innovation, manufacturing, and business. It was a time of rising wages and better access to housing and goods that had been out of reach for most Americans during the Great Depression and World War II. The industrial giants formed in the Corporate Era, such as General Electric (GE) and General Motors (GM), had joined the war effort and scaled up production for the military in the 1940s. By the end of the war, years of federal contracts had created close relations between government and corporations, forming the military industrial complex that would shape business and government in the Cold War era. Charles E. Wilson (1890–1961), for instance, made an easy transition from head of GM in 1953 to Secretary of Defense in the Eisenhower administration, returning to the leadership of the automotive giant a few years later.

Manufacturers retooled their factories quickly after the war to serve consumer markets both at home and in Europe. American factories became the cornucopia of postwar abundance. In 1945, Americans made 70,000 cars. Ten years later, there were more than 8 million cars rolling out of factories and into dealerships and newly constructed suburban garages. GM, the largest business in the United States in the 1950s, churned out appliances, such as Frigidaire refrigerators, that formed the material basis of the American dream and filled the kitchens of new, mass-produced suburban houses. Competitor, GE made aircraft engines, but also Hotpoint refrigerators and a range of small appliances from percolators to toasters.

As with the early era of manufacturing, big business could produce more products than people could buy, and they relied on advertising to help spur consumer desire. From GM's appliance-filled film *Design for Dreaming* (1956) to the "Live Better Electrically" campaign sponsored by power companies and appliance makers, advertisers helped stoke demand. For their part, civic and business organizations, like the National Association of Manufacturers (NAM), celebrated the power of American manufacturing as the engine of prosperity. NAM's television series, *Industry on Parade*, which first aired in 1950, covered industry big and small in hundreds of short episodes that ran for a decade on network television.

New kinds of credit enabled millions of Americans to buy single-family houses and fill them with a host of new "populuxe" goods designed for the exuberant consumer market. It was a period in which more Americans, including married women, entered the ranks of office workers. Along with the manufacturing sector, white-collar work grew in this period, producing a wide middle class of professionals and managers. This growth in white-collar work was not without its critics, however. Sociologists, journalists, and cultural critics studied this new generation of office workers and managers and worried about shallow materialism and social conformity, represented by the iconic gray flannel-suited organization man, that would deplete American individualism and possibly sap American business.

As productivity increased, so did rates of unionization; a blue-collar middle class came of age. Despite the passage of the Taft-Hartley Act (1947) and the first wave of right-to-work laws designed to chip away at labor rights created in the 1930s, these workers enjoyed wages and benefits that gave them a greater stake in a consumer society. With the merger of the American Federation of Labor (AFL)

above, left: **AFL-CIO first joint convention, December 5, 1955.** George Meany, president of the American Federation of Labor (left) and Walter Reuther, president of the Congress of Industrial Organizations (right), celebrated the merger of their organizations at the first joint convention in New York City.

above, right: **AFL-CIO convention badge, 1955.** The American Federation of Labor (AFL) and the Congress of Industrial Organizations (CIO) commemorated their merger in 1955 with a special convention badge that symbolized solidarity, using the clasped hands, and the location of the meeting in New York City.

and the Congress of Industrial Organizations (CIO) in 1955, organized labor reached the apex of its power; almost a third of American workers were union members.

However, American prosperity was not shared equally, and journalists and economists debated its sustainability and the role of business in American society. In his best-selling book, *The Affluent Society*, economist John Kenneth Galbraith worried that Americans spent too much money on consumer goods and neglected the common good. He asked policymakers to commit some of the vast resources of the United States to schools, healthcare, and social services that could help more citizens share in the abundance created by American capitalism. He also warned readers about the consequences of not investing in social infrastructure—a future where gains in wealth went to an increasingly small group of people and where the nation neglected investing in the common good.

Some Americans, especially those who remained in cities and rural areas, or whose encounters in the marketplace were shaped by racial segregation and discrimination, did not experience the same prosperity. The Civil Rights movement, President Lyndon Johnson's War on Poverty, and the Poor People's Campaign of 1968 all sought to make inequality visible and find solutions to these persistent problems. But these issues were complicated by the slowdown in productivity, a leveling off of wages, and the energy crisis of the 1970s. When President Jimmy Carter took office in 1977, he confronted a very different economic landscape from those of his predecessors. The energy crisis and what economists and journalists termed stagflation—stagnating wages and growing inflation—led to a questioning of the promises of consumer capitalism that had defined the beginning of the era.

opposite: **Marlow Family, 1961.** In 1964, President Lyndon Johnson declared a "War on Poverty," citing the abysmal conditions of Appalachian families. Lady Bird Johnson selected images of the Thomas Marlow family to help publicize the campaign. Congress soon passed legislation that led to Head Start, Job Corps, Medicare, and Medicaid.

INNOVATIONS IN BUSINESS: IBM

Since the emergence of the market revolution in the eighteenth century, information has been the "lifeblood of business" (Lubar 1993). Businesses, from early merchants to large corporations, used various technologies—accounting systems, typewriters, filing cabinets, adding machines—to control and organize everything from profits and debts to payroll and inventory. But as America moved into the consumer era, selling technology to manage information became a big business unto itself. No company was more successful in this field than the International Business Machines Corporation— IBM.

The company that would become IBM began in the work of Herman Hollerith, inventor of a punched card tabulating machine that was first used to process the 1890 census. He created the Tabulating Machine Company to market his new device. In 1911, this firm merged with two others to form the Computing Tabulating Recording Company. It sold a wide range of products, including time-keeping systems, scales, automated meat slicers, and, of course, punched card tabulators. After struggling to unify the combined business, the company looked for new leadership. In 1914, it hired Thomas J. Watson, Sr., from the National Cash Register Company as general manager. Within a year, he was president, and began revamping the business, giving it a distinctive culture and a broader vision.

Watson believed strongly in employee education and self-motivation. He introduced the slogan "THINK," which became the company's motto and symbol. "The trouble with every one of us is that we don't think enough," Watson would say. "Knowledge is the result of thought, and thought is the keynote of success in this business or any business" (IBM Archives). He also insisted that IBM employees respect both themselves and their customers, dress professionally, and give excellent customer service.

GORDON E. MOORE
1929–Present

ROBERT N. NOYCE
1927–1990

Gordon Moore and Robert Noyce founded one of the most successful information age businesses, Intel Corporation, in 1968. Moore is best remembered for "Moore's law" (1965), which predicted that the number of components that could be packed onto microchips would double every two years. His law has proved remarkably accurate, due to aggressive innovation. Robert Noyce co-invented the integrated circuit with Jack Kilby, which made possible personal computers and many other electronic devices. "Innovation is everything," Noyce once said. "When you're on the forefront, you can see what the next innovation needs to be. When you're behind, you have to spend your energy catching up" (Botkin et al 1984).

Gordon E. Moore, 2005.   Robert N. Noyce, circa 1990.

In 1924, Watson renamed the company International Business Machines, and focused it on developing punched card systems that could serve a wide range of uses. Like other firms, it struggled during the Depression, but in the mid-1930s, it landed America's biggest bookkeeping job: implementing the new Social Security system. This massive assignment required maintaining and processing records on the earnings of millions of Americans. Nothing on this scale had ever been attempted. Based on success with the project, IBM grew dramatically, both domestically and internationally, and became a world leader in information processing.

In 1952, Thomas Watson, Jr., succeeded his father as president of IBM. For the company to retain its leadership position, Watson, Jr., believed it had to enter the new field of digital computers. His father and others at IBM were not convinced. Although in hindsight their hesitance seems foolish, at the time they had good reason to be cautious.

The computer industry in America began with ENIAC, the Electronic Numerical Integrator And Computer, which the Army built in World War II to compute ballistics tables. It had proven that digital computers could be practical. But it also required 18,000 vacuum tubes, and had to be re-cabled for every job. It was far from being a commercial product. In the postwar years, a number of companies jumped into the computer field, but because electronic processing and memory technology were immature, they struggled to make reliable machines that were profitable. Most potential customers were content to stick with their familiar and reliable punched cards, IBM's core business.

Watson, Jr., decided to forge ahead, believing IBM could lead its large customer base into the future. IBM's first computer, model 701, debuted in 1952. It was designed primarily for scientific processing. A year later, IBM introduced a computer designed primarily for business, model 650. It was the first mass-produced digital computer, and was soon adopted by large banks, insurance companies, and scientific laboratories. IBM chose to lease, rather than sell the machines, limiting its customers' investment risk. This strategy also let the company maintain control over the machines and negotiate lucrative service contracts. Like ENIAC, the 650 still relied on vacuum tubes, but it had far fewer and they were more reliable. To manage storage, IBM introduced an innovative magnetic drum, which could hold some 20,000 digits—a major achievement in this era. Although the 650 became the most popular computer of the 1950s, it was not right for all customers, and IBM developed a number of additional models. By 1960, it was dominating the business computer market, just as it had mechanical data processing in the 1930s and 1940s.

As Watson, Jr., watched the computer market grow, he became convinced that IBM had to take another bold step. Instead of continuing to develop a variety of special-purpose machines, customized for different types of users, he believed IBM should create a single, integrated system of computers that met all customers' needs. This was an enormously complex and expensive project, costing $5 billion, when annual company profits were only half that amount. In essence, he bet the company. And won.

IBM called its new product System/360, because it could meet the entire circle of customer needs. In addition to being an integrated family of computers, which required major advances in software development, System/360 included dramatic hardware advances, such as transistorized modules called solid logic technology. Announcing the system in April 1964, Watson said:

> System/360 is a single system spanning the performance range of virtually all current IBM computers… It is the balancing of these factors—all available within a single system using one set of programming instructions—that will make it possible for a user to select a configuration suited to his own requirements for both commercial and scientific computing. A user can expand his System/360 to any point in its performance range, without reprogramming (IBM Archives).

ADVERTISING CREATES A MARKET FOR PERSONAL COMPUTERS

With the advent of personal computers, companies first advertised to hobbyists, then to a more general market that included families and schools. Altair offered one of the first computer kits in 1975, designed around an Intel chip, and available to the hobbyist market. By 1977, a host of companies, including RadioShack, made personal computers and used advertising to create a market for them. Ads made the case that families could use computers for everything from managing household accounts to entertainment. In 1980, RadioShack ran this cross-promotional advertisement with DC Comics to reach the youth market.

"The Computers That Saved Metropolis!" Superman, July 1, 1980.

above, top: **IBM System/360.** IBM introduced the System/360, a mainframe computer in 1964. The system referred to a "family" of components that could be rented by large-scale users, including the federal government, banks, and insurance companies, tailored to their specific needs.

above, bottom: **Model of IBM System/360, about 1964.** Small-scale models of the System/360 allowed IBM salesmen to demonstrate the flexible components, including central processing units (CPUs), data storage units, keyboard, and printer to potential buyers. Models were branded in the company's trademarked orange.

IBM sold thousands of System/360 computers, and greatly increased its lead in the computer field. Its rivals, including NCR (formerly National Cash Register), Honeywell, and General Electric, now were derisively called "the seven dwarfs," and struggled to find profitable areas where they could still succeed. Computers became major fixtures in American business. Some worried that automation would mean unemployment for thousands of clerks and accountants, but the change also created new jobs: for computer engineers, software programmers, and data entry staff. Whatever the social cost, implementing computer systems became increasingly necessary for businesses to remain competitive. This new age of electronic speed and efficiency would soon ripple through many other areas of business and manufacture.

IBM's business success spotlighted its distinctive corporate culture. The company's service technicians were required to wear white-collared shirts and ties, even when working on the machines. THINK became the title of the corporate magazine, and appeared on notepads on which employees could jot down their innovative thoughts. The one-word mandate was translated into multiple languages for desk signs, spreading the corporate culture of individual innovation beyond the headquarters in Poughkeepsie, New York.

As dark-suited, white-collared IBM salesmen traveled to cities as far-flung as Venice and Berlin, armed with models of their equipment, the THINK sign was translated into each of the seventy-nine languages of the IBM offices around the world by 1950.

IBM's organizational branding extended to the signature colors used on its machines and its distinctive blue logo. The System/360 sported IBM's trademarked blue and orange colors and was easily recognizable at a glance. It came to epitomize American business culture and global influence in the Consumer Era.

IBM salesman in Venice, 1966. IBM Archives. IBM was a global company and its salesmen traversed the globe after World War II. A typical white-collared dark-suited employee commuted to work in a picturesque gondola, an image of American business power abroad and proof of the company's global reach.

PETER F. DRUCKER
1909–2005

Peter Drucker earned the title "the man who invented management" for his prolific writings on management theory and business organization. A professor at New York University, he published some thirty-nine books in the field over several decades. Influential in describing business practices, he coined the term "knowledge worker" and predicted the shift from an industrial production to an information society. He believed that managers, as an influential group, not only had a responsibility to business but also to society. He wrote in 1974 that "If the managers of our major institutions, and especially of business, do not take responsibility for the common good, no one else can or will" (Drucker 1974).

Peter F. Drucker, 1960s.

Group of THINK objects. "Think," declared Thomas Watson, Sr., CEO of IBM in the 1920s, and the mandate stuck, becoming the company's long-lasting slogan. Desk signs, notepads, and the title of the company magazine all reminded IBM employees of the importance of innovation.

FRANCHISING: BUSINESS INNOVATION BIG AND SMALL

With a thriving consumer economy, many businesses looked for opportunities to reach buyers in new ways and in new places, such as the suburbs and along the developing interstate highway system. Beginning in the 1950s, more small entrepreneurs turned to franchising as a way to grow their businesses. The concept of franchising—businesses working through local affiliates or outlets to sell their products—began in the nineteenth century. But the business form increased dramatically in the Consumer Era around service industries, with the rise of the highway system, changes in work, lifestyle, and more available credit.

Franchising also offered a way for Americans with little capital or business experience to take a limited risk and become proprietors of their own businesses. The franchise system gave these small investors access to a standardized system, discounted goods, and national advertising as seen in the national trade journal, *Modern Franchising*, which touted the successes of average people who followed their dreams and became franchisees. Upward-pointing arrows adorned the cover of each issue, illustrating the possibilities for financial growth. The journal also ran ads that invited readers to take the leap and buy into a franchise and many would-be entrepreneurs took loans to open fast-food restaurants, hotels, and gas stations of their own. As historian Tom McCraw has noted, franchising grew more quickly than any other form of business in the Consumer Era and exemplified new trends in the decentralization of decision-making and consumer empowerment. In addition, the nature of franchising changed in the postwar era, concentrating on services rather than selling products made by a large company, as in the case of car dealerships. Franchises were also a good deal for the parent companies, because while the individual proprietors, or franchisees, often paid

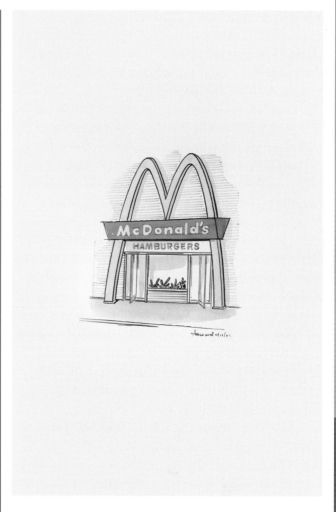

Concept drawings for McDonald's franchises, 1961–1979. Graphic artist and adman Arthur Bernie Wood founded Admart, Inc. Advertising in the early 1960s and created a unified look for McDonald's franchises, from the architecture to the packaging, advertising, and fun premiums for kids.

for land, construction costs, and supplies, they had to adhere to certain standards for architecture, food, and service set by the company.

Many of the America's leading franchises started in the 1950s as small ventures and grew with suburbanization, increased automobility, and changing patterns of work. Two of the most well-known, McDonald's and Kentucky Fried Chicken, were shaped, in part, from innovations by franchisees. In 1954, when Ray Kroc, a middle-aged salesman, became the franchise agent for the McDonald brothers and their hamburger stands, franchising was a system in which parent companies sold the rights to their outlets. Most businessmen who started franchises hoped to make money from these one-time sales rather than the continuing management of outlets.

Kroc, on the other hand, saw the potential of licensing McDonald's franchises and keeping all the affiliates under the organizational umbrella of a larger corporation. He set up a system where franchisees paid a one-time fee for the franchise plus an annual royalty based on a percentage of gross sales. Most of the annual royalty fee went to Kroc's new company, McDonald's Corporation, established in 1955, and he paid for all the expenses of overseeing and supplying the franchisees. Although Kroc did not see a profit until the early 1960s, the innovative system helped establish McDonald's as a national business. Kroc emphasized that McDonald's needed to give its franchisees the tools (from systems to advertising) to be successful, and a franchisee's success would only increase the wealth of the larger company. He reportedly coined the slogan, "In business for yourself, but not by yourself" (McCraw 2009). Although it took time to grow the company based on the expansion of franchisees and the small royalty payments, eventually it worked: McDonald's Corporation became one of the most successful franchise businesses of the twentieth century.

One of the keys to Kroc's success lay in balancing centralized control with decentralized decision-making. The parent company sourced supplies, provided national advertising (paid for in part by franchisees), designed the brand, and owned the land. The parent company also required franchisees to adhere to Kroc's formula for success with consumers, known as the "QSC," or quality, service, and cleanliness. Operators maintained their independence to solve problems on the ground, and controlled hiring, pay, and pricing. Franchisees also invented new menu items that then became part of the national offerings, such as the Big Mac, and, along with suppliers, found ways to make operations more efficient. This flexibility gave McDonald's a competitive edge and the company grew into a global empire, opening outlets in over 120 countries by the twenty-first century. McDonald's success mirrored the more general success of franchising as a business model in the latter half of the twentieth century. According to a 1990 study of the five-year survival rate of small business in the United States, franchises had a greater chance of success than nonfranchised businesses.

At the same time that Ray Kroc began working for McDonald's, Harland Sanders turned his secret recipe for pressure-fried chicken into a franchise business. Sanders brought a good deal of experience to franchising. He had run a roadside café and tourist court in Corbin, Kentucky, starting in the 1930s, where he served a Southern menu of steaks, fried chicken, biscuits, and country ham. The café even merited a mention in food writer Duncan Hines's *Adventures in Good Eating*. In the 1950s, the new interstate highway system bypassed his business, reducing the flow of tourists and forcing him to sell. He then turned to a franchise model. First he patented his cooking method and devised a way to get three meals out of one chicken. Then Sanders traveled the country, demonstrating his frying technique and promoting the franchise opportunities. He leased his pressure cookers, supplied the secret but standardized seasonings, and got a return on each chicken sold. But scaling up to a national business relied on the partnership, financial support, and innovations of franchisees.

It was one of his first franchisees, a man named Leon "Pete" Harman in Salt Lake City, Utah, who helped Sanders create Kentucky Fried Chicken as a stand-alone business and build a model for the franchisees. Harman dedicated his restaurant to Sanders's fried chicken, created the take-out bucket, thought up the catchy slogan, "It's finger lickin' good," and scaled up the frying of chickens to mass proportions. And, like Ray Kroc and McDonald's, Harman did away with the sit-down dinner model; he realized that in the increasingly mobile society, Americans wanted to take lunch or dinner on the road with them.

own a franchise in america's future

modern FRANCHISING

January-February, 1965
Thirty-Five Cents

Membership List — IFA

Make People Want

Franchise City — — Fairfield

A New Husband Task Force

Cover of *Modern Franchising*, January/February, 1965. Everything about *Modern Franchising*, a trade journal serving franchise owners, linked franchising to the idea of individual opportunity. The upward-pointing arrows on its cover spoke of future profits. Enticing advertisements and success stories encouraged readers to be their own boss.

For his part, Sanders cultivated his persona as the white-suited, ivory-haired gentleman. Sanders had received the honorary commission of "colonel" from the Kentucky governor. The title was ceremonial, but, with the keen sense of a good salesman, he turned it into a character that gave his product an identity. Even before he closed his Corbin café, he adopted the trappings of a Southern gentleman. As he began the franchise business, he developed the neat suit and string tie, turning the guise into a powerful brand that gave the restaurants a nostalgic identity that consumers associated not only with the South but also with home cooking. To make his sales pitch to potential franchisees, Sanders also carried photographs of Harman's restaurant, with crowds lined up outside, with him as he signed up new franchisees. The white suit and good manners were more than a branding strategy. Sanders, a committed Rotarian, believed that businesses should serve communities. He demanded cleanliness, friendliness, and quality food at his outlets, and worked on an honor system sealed with a handshake.

Sanders and Harman tapped into a desire among consumers for convenience food and met a need for meals that mimicked home cooking at a time when more women—wives and mothers— were working outside of the home and had less time to cook. The franchise grew quickly. With the business expanding to all points on the weathervane, Sanders sold out in 1965 for $2 million. In his own words, he wasn't an "organization man" (Sanders 2012). Sanders would remain the iconic image of the brand, but would later criticize the deterioration of the product as just a shadow of his original recipe.

Colonel Sanders weathervane, 1950s–1960s. Weathervanes, in the image of Kentucky Fried Chicken's founder and self-styled brand icon, symbolized the reach of the franchise network to all points of the compass. They also lent an old-time feel to the thoroughly-modern architecture of the drive-in restaurants.

THE BUSINESS OF BROADCASTING

In the Consumer Era, television became a big business, from broadcasting to making and selling television sets. Although inventors developed the technology before World War II, television remained too experimental and expensive for most Americans until the postwar years. Nevertheless, the established broadcasters including NBC, led by David Sarnoff, and CBS, along with manufacturers of receivers, and even local entrepreneurs who ran stations, saw television as a business opportunity. Relative newcomers, including Allen B. Du Mont, competed for broadcast licenses and affiliate stations, advertising sponsors, and for audiences. An engineer and inventor, who had been experimenting with the technology of television since the 1930s, Allen Du Mont briefly challenged the supremacy of established networks NBC and CBS by starting his own eponymous network, DuMont (the network used a slightly different spelling of his name, omitting the space). Allen Du Mont entered the field on two fronts: he manufactured a premium line of television receivers based on his pioneering technology, and he established what would become an innovative, if short-lived, fourth network.

For their part, consumers eagerly bought televisions and tuned in to whatever programming was available, creating a ready market for sets, shows, and advertising. Between 1950 and 1955 the number of sets sold rose from about 3.1 million to 32 million. By 1955 televisions sets were a must-have technology and were flying out of showrooms at the rate of 10,000 per day. Within ten years, by 1960, TV had become a common household appliance; almost 90 percent of American homes had sets and they were turned on for about five hours a day (Pursell 2007). American families not only watched TV, they made room for the technology in their homes, giving the bulky sets a place of honor in the center of their living rooms, arranging their furniture around them, and buying a host of new accessories and furnishings to make life with television more comfortable. A constellation of new consumer products, from lamps that cast ambient light to reduce eye-strain to TV trays, so families could eat dinner in front to the tube, appeared in department stores and catalogs.

Allen Du Mont earned the title of "father of television" for pioneering the cathode ray tubes that made the transmission of images possible in the late 1930s. He began manufacturing and selling television receivers near his research laboratory in New Jersey in 1938, just before the public debut of the new technology at the 1939 New York World's Fair. He was not alone in anticipating a new market; radio manufacturers including General Electric, Philco, and RCA (which owned the National Broadcasting Company, NBC)

Allen B. Du Mont with cathode ray tube, 1940s. The cathode ray tube, also known as "the magic eye," transmitted pictures onto television screens. Du Mont also manufactured television sets and started his own television network.

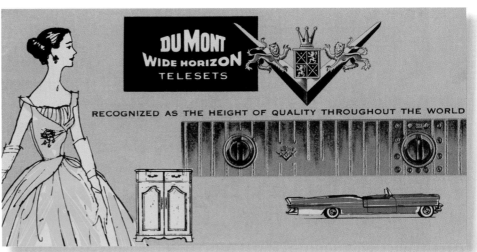

above, top: **DuMont sales banner and product literature, 1950s.** Salesmen used banners to attract customers and convey the high quality of DuMont televisions. Sales literature explained DuMont's innovative technology and told consumers how broadcasting worked.

above, bottom: **DuMont "Revere" entertainment center, 1947.** This DuMont television and entertainment center included a radio receiver with high fidelity amplifier and record changer. The cabinetry for television sets imitated popular furniture styles. This DuMont Revere model enclosed modern technology in colonial revival style.

anticipated a future market in television and began developing receivers and laying the groundwork for regional and national broadcasting. At a moment when relatively few Americans owned television sets, the DuMont company marketed to an upper-middle-class consumer with advertising that associated its product with other luxury goods. Its receivers became known as the Cadillac of television sets; they were expensive, and they mimicked popular furniture styles. In the late 1940s and early 1950s, DuMont offered colonial revival designs that drew on early-American lines and details. Fine wood cabinets with molding and brass knobs conveyed a sense of tradition, class, and solidity.

Allen Du Mont, however, was not satisfied with just manufacturing receivers. He also waded into the heady waters of the broadcasting business. In the 1940s, he got a permit for an experimental station in New York City and opened a small studio to produce programming on Madison Avenue. Also in these early years, he entered into a partnership with Paramount Studios in Hollywood. Paramount was interested in transmission technology and wanted to produce programming, although broadcasting during World War II was limited. Although he entered television broadcasting early, Du Mont's broadcasting business was undercapitalized compared to NBC and CBS and at a disadvantage in terms of the number of affiliate stations he could draw into his network and who would help pay for programming.

Du Mont opened a second station in Washington, D.C. in 1945, but the Federal Communications Commission (FCC) ban on new station licenses tempered his success in constructing a viable network. The FCC, the chief regulator of radio, telephone, and television, also played an important role in determining the shape and future of television broadcasting. It was chiefly responsible for allocating new stations and the frequencies those stations used. Between 1948 and 1952, the FCC froze the process for applying for new broadcast licenses, as the commission internally debated its policies for allocating stations. The FCC limited the number of channels operating in VHF and assigned newcomers, including DuMont, UHF channels, which required most television owners to buy converter boxes so they could receive the broadcasts. The "freeze" and subsequent limitation of VHF channels gave giants like NBC and CBS, which had established networks during the radio era of the 1930s, a prime position after the 1952 freeze was lifted and curbed serious competition from other networks. Any broadcaster that wished to be competitive at the national level had to build a network of local affiliate stations, and the FCC limited the number of new local stations.

By 1950, the production side of Du Mont's studio had an experienced and visionary team of managers, directors, writers, and talent. Working with small budgets and limited crew, the DuMont network created models for advertising and programming that shaped the future of television. As with radio, commercial sponsors paid for the majority of programs on early TV. The DuMont network had no legacy relationships with sponsors, unlike NBC and CBS, and had to innovate. DuMont courted advertisers with modest budgets and sold advertising time on programs to multiple sponsors, initiating a move away from the one-sponsor-one-program format. DuMont also allowed sponsors to target advertising by airing commercials only on certain affiliate stations. DuMont producers experimented with daytime programming, an area that the television broadcasters with roots in radio were slow to enter. There was a small market for daytime programming and DuMont produced a morning talk show, a home shopping program, and a children's program titled *Your Television Babysitter*, which offered content for housewives, mothers, and children. These innovative shows attracted advertising sponsors and created much needed profit for the network. In addition to taking advantage of underused time slots, DuMont produced some critically acclaimed programming and launched stars like Jackie Gleason for prime time.

The DuMont network's main competition came from the other newcomer, ABC. In the early days

"Telebugeye" cartoon, by Munro Leaf, *Ladies Home Journal*, 1950. *Ladies Home Journal* warned mothers that too much television could be bad for their children's health. The cartoon illustrated the potential ill effects of television on kids—poor posture, weak eyes, and laziness—and captured the widespread debate about the social value of television.

of television broadcasting, in the 1940s and early 1950s, ABC (formed from the FCC's forced break-up of NBC) and the DuMont network vied for third place among national broadcasters. Although he seemed to be in an almost perfect position to win that spot, Allen Du Mont and his staff struggled to find affiliate stations for their programs and advertisers who would pay for programming. Senior managers at DuMont also struggled for control over the company with their partners and stockholders at Paramount. ABC formed a lucrative alliance with Walt Disney studios in 1953 that provided financing as well as weekly programming for the station. DuMont's deal with Paramount wasn't so sweet. In fact, leadership at Paramount gained control of the business through stock options and eventually removed Du Mont from his position as president in 1955. Du Mont lost the business he built and broadcasting shrunk to three major networks. When many Americans turned on their television sets after

1955, their choices would include programming created by local stations and prime-time offerings from the three big networks, NBC, CBS, and ABC.

Although Americans bought televisions in record numbers, placed them in the center of their living rooms, and believed that television could bring the family together, they also worried about the relationship between television and the common good. Television came with its own set of cultural discontents, including a worry that that too much television could have detrimental effects on American children: ruining their eyes, making them passive vessels for potentially dangerous ideas, creating bad habits, and even leading to delinquency if not regulated by parents and, at a higher level, the federal government. Television, because it used public airwaves, was regulated by the FCC, which not only issued licenses for new stations but also monitored programming. Those frequencies were also public spaces to be used for the common good, not just private gain, and television programming, especially programming aimed at kids, came under tough scrutiny by parents' associations, journalists, politicians, and even broadcasters themselves.

Allen Du Mont, for instance, saw great potential for television as a "public service," but also remarked frequently that he was disappointed in the low-quality programming and violence on television. Perhaps most famously, on May 9, 1961, head of the FCC, Newton Minnow addressed the National Association of Broadcasters and characterized broadcasting as "a vast wasteland," prompting a reexamination of the relationship between broadcasting and the common good.

This national conversation about the social and educational role of television resulted in the creation of the Corporation for Public Broadcasting (CPB), through an act of Congress in 1967, and the Public Broadcasting System (PBS) in 1969. Although some noncommercial children's and educational programming existed in the 1950s, it was limited. The Ford Foundation funded the creation of National Education Television (NET) in 1952 to provide quality educational programs for primarily

ADVERTISING AND CHILDREN

Postwar television targeted children as one of the most promising groups of consumers. Children's programming mixed seamlessly with advertising and ran the gamut from westerns to variety shows; it was designed not only to entertain but also to sell products. Of all the pitchmen on TV, Howdy Doody was perhaps the most successful. NBC, the creator of the *Howdy Doody* show, used the character to sell advertising sponsorships, to license products (dolls, puppets, books, clothing, records, stationery, adhesive bandages, and tie-ins with cereal), and to promote color television sets introduced by NBC's parent company, the Radio Corporation of America (RCA).

Howdy Doody puppet, circa 1950s.

Cookie Monster sweater, 1970s–1980s. Although it aired on public television, *Sesame Street* had its commercial side. The Children's Television Workshop, the nonprofit company behind *Sesame Street*, licensed images of its characters and name in the 1970s to pay for future productions. J.C. Penney was among the first retailers to sell *Sesame Street* licensed clothing.

adult audiences but also included some offerings for children. In addition, some university- and community-owned stations developed educational programs for children. But it was the formation of the CPB that channeled federal funding for public radio and television broadcasters, and the founding of PBS, that brought many of these efforts together. PBS stations also provided an outlet for one of the most successful and beloved children's shows of all time, *Sesame Street*. The brainchild of the Children's Television Workshop and informed by research on early-childhood learning, *Sesame Street* provided a model for educational television programming aimed at preschool children. The one-hour program that first aired in 1969 addressed a disparity in preparation for school among lower-income children, but was watched and widely admired by a large swath of children and parents. Despite the success of *Sesame Street*, debates about the quality and social value of television and the influence of commercial interests on programming aimed at children would continue.

Although the big three, NBC, CBS, and ABC, dominated national programming after 1955, local stations provided opportunities for entrepreneurs to shape broadcasting and serve the common good. In particular, the development of Spanish-language television in the United States focused on underserved audiences and used television to not only entertain, but also to address civic and social issues. In the postwar period, broadcasters in Puerto Rico, Texas, New York, and California built

Spanish-language television stations that succeeded where the DuMont network had failed in creating a fourth network. The two largest Spanish-language networks of the late twentieth century, Telemundo and Univisión, had their roots in the Consumer Era. These media pioneers cultivated new markets for advertisers and developed programming that spoke to issues local audiences cared about.

Spanish-language television grew from radio broadcasting and had its roots in the beginnings of broadcast media. Often the entrepreneurs who launched stations in the Consumer Era had experience in radio. Ángel Ramos, media entrepreneur, owned Puerto Rico's main newspaper, *El Mundo*, and radio stations before launching the WKAQ-TV in 1954. Known for its locally-produced soap operas or *telenovelas*, the station provided the starting point for what became the Telemundo network. San Antonio's KCOR-TV, established in 1955, grew out of the successful Spanish-language radio station founded by businessman, president of the League of United Latin American Citizens (LULAC), and community advocate Raoul A. Cortez. KCOR-TV, the first Hispanic-owned, full-time station in the continental United States, aired a vibrant mix of programming, including: locally-produced variety shows, movies, news, interviews with politicians, and editorials that voiced community concerns.

Cortez eventually sold the station in 1961 to his son-in-law Emilio Nicolas, Sr. and a group of investors including Emilio Azcárraga Vidaurreta, who owed Mexico's main television network. Nicolas, Sr. and his partners changed the call letters of the San Antonio station to KWEX, worked to find sponsors, bought affiliate stations, and built the Spanish International Network (SIN). By the early 1970s, SIN spanned the United States from New York to Los Angeles, serving an under-recognized market of Spanish-speaking Americans and convincing advertisers of their buying power. With a subsequent sale to Hallmark Cards in the 1980s, the network became Univisión.

RAOUL A. CORTEZ
1905–1971

Raoul Cortez thought media should serve the community. A Mexican immigrant, he sold eggs and worked as a sales representative for a local brewery in San Antonio, Texas. His media career began at the Spanish-language newspaper, *La Prensa*. From there he opened a radio station, KCOR, in 1946, where he broadcast in Spanish and addressed the needs of the Mexican American and immigrant communities. The station became the voice of the Spanish-speaking community in South Texas and used the phrase "la voz Mexicana en San Antonio" as its station identification. Cortez served as national president for the League of United Latin American Citizens (LULAC) and advocated for the desegregation of Texas schools and the fair treatment of immigrant workers. He eventually expanded his broadcasting business to television in 1955.

Portrait of Raoul A. Cortez, 1940s.

NEW WAYS TO PAY: CONSUMER CREDIT

"We are in a consumer credit explosion–a revolution in modern banking," wrote Garrison A. Southard, Jr., executive manager of the California Bankcard Association in 1967 (Zumello 2011). And he was right. With innovations in consumer lending in the postwar period, commercial banks developed the universal credit card that offered Americans new ways to finance their purchasing. The universal credit card changed how Americans thought about credit and debt and transformed how banks interacted with consumers.

American consumers were no strangers to credit, and what Southard and other commentators observed was an expansion of new forms of consumer finance. Credit had a long history in the United States, dating to the market revolution of the eighteenth and early nineteenth centuries when merchants regularly extended credit to everyone from farmers buying seed to housewives purchasing staples. In the early republic, especially, credit greased the wheels of business and played an essential role in keeping individuals, businesses, and the nation solvent.

Debt on account books evolved into store credit, issued by individual shopkeepers and later formalized and rationalized with the rise of big business, including department stores and later General Motors. Department stores were among the first to offer deferred payment plans like layaway, and eventually, revolving accounts attached to charge plates. In 1919, General Motors Corporation offered a new financing program for its automobiles, General Motors Acceptance Corporation (GMAC). With financing available on the spot, prospective buyers could afford a new car, or even a second car at the same time as the manufacturer took steps to ensure a steady demand for its increasing supply of automobiles. The early installment buying plan gave customers up to twelve months to pay off their new automobiles. The GM financing model was widely adopted (except by Ford) and by 1922 there were over 110 auto finance companies and more than 70 percent of automobiles were sold on time.

Diners' Club, an elite circle of businessmen, issued its first charge card in 1950. A precursor to universal bank-issued credit cards, the Diners' Club card and similar American Express card initiated charge accounts that could be used at participating restaurants, clubs, and stores. Replacing cash with a credit card allowed cardholders to avoid the embarrassment of not having enough cash in social situations. Alfred S. Bloomingdale, grandson of

Washington Shopping Plate, circa 1960s. This card allowed holders to charge purchases at seven different stores that made up the retailers' association that had issued the plate. Made of durable plastic, it fit easily into a wallet and listed participating stores and terms of credit on the back.

Alfred Bloomingdale's Diners' Club credit card, 1958. Diners' Club issued these paper cards to a select group of club members in the 1950s. Rather than carrying cash, businessmen used the cards to charge dining and entertainment expenses. Bloomingdale served as the long-time president and later chairman of the board.

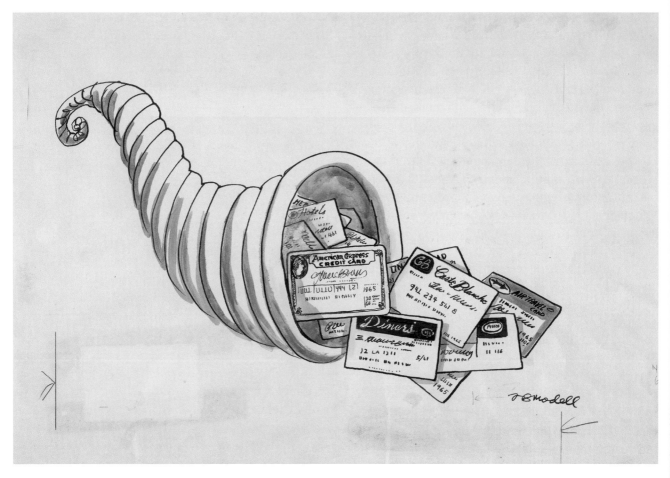

Credit card cornucopia drawing, by Frank B. Modell, circa 1964. *New Yorker* artist Frank B. Modell characterized the explosion of credit cards available to Americans in the 1960s and associated them with abundance. Alfred Bloomingdale acquired the original artwork from Modell for his collection of credit card cartoon art.

BankAmericard, 1978. In 1958, Bank of America launched its BankAmericard, which would become VISA. Although the process of working out this new kind of credit was fraught, ultimately, the bank-issued card was a success because of the flexibility it offered consumers.

the department store baron Lyman G. Bloomingdale, started Dine and Sign, a similar concept, which merged with Diners' Club in 1951. Bloomingdale served as chairman of the board of Diners' Club from 1964 until 1969.

With the success of department store credit cards and the Diners' Club and American Express cards, commercial banks seized an opportunity to sell credit by offering cards attached to revolving accounts. These cards were "universal" in the sense that they could be used anywhere, including the department store, gas station, restaurants, hotels, consolidating credit under one account and offering an incredible opportunity for commercial banks to make money by collecting interest on the accounts. Previously, stores and clubs that issued charge cards, where the balance was paid at the end of the month, did not collect interest, but the innovative new credit cards did. Bank of America issued its BankAmericard (what would become VISA) in 1958 and Interbank launched MasterCard in 1966. New credit options followed these initial cards and, with them, increasing consumer debt. By the late 1960s, revolving credit plans offered by commercial banks and that allowed borrowers to carry the debt from month to month without paying off the balance, increased 200 percent, prompting an analyst for Salomon Brothers to note that credit cards were "becoming as common as miniskirts" (Zumello 2011). Not only did credit cards become common, they became networked. Evolving alongside new computer systems, magnetic strips on the back of credit cards, first developed by IBM in 1969, gave banks and retailers more control over accounts and data about consumers.

By the late 1960s, millions of American wallets held durable, plastic credit cards with the name of a bank on the front and terms of service on the back. But as credit became more available and easier to use, Americans wrestled with the meaning of the credit card: who could and couldn't use it, and how it should be used. Cartoonists worked out these questions in the public forums of major newspapers and magazines. The cartoons showed Americans struggling with the appropriate uses of credit cards. The illustrators overwhelmingly depicted men as the primary users. But even male consumers struggled with what counted as appropriate use of a "universal" credit card that could theoretically be used for anything. Cartoonists also captured lingering doubts about getting into debt. Even if debt had a long history and credit in the postwar years became a right associated with full citizenship in the evolving consumer economy, many Americans still worried about the implications of personal indebtedness. One of the most compelling illustrations, "467-821," followed an enthusiastic user of credit all the way to prison.

across from top left: **"467-821," by Ted Key, circa 1960.** Ted Key, the artist behind the famous *Hazel* comic strip, drew this darkly humorous series of cartoons in the early 1960s. As consumers struggled with the notion of easy credit, cartoonists warned of overspending.

"467-821."

"....GET RIGHT TO WORK ON THE ALTERATIONS. NUMBER - 467-821 ..."

"NICE TO HAVE YOU WITH US, SIR. NUMBER 467-821."

"NUMBER 467-821. RIGHT, SIR."

"THANK YOU, SIR. 467-821."

WILLIAM "BILL" BERNBACH
1911–1982

Tired of formulaic advertising, William Bernbach partnered with Ned Doyle and Maxwell Dane, forming Doyle Dane Bernbach (DDB) in New York City in 1949. Their work ignited a creative revolution: DDB ads used artistry and quirky copy to sell. Bernbach once said that "Nobody counts the number of ads you run; they just remember the impression you make" (DDB, Why Bernbach Matters). DDB's ad changed impressions of Orhbach's department store. In this premier example of creative advertising, DDB used haute couture and catty copy to remake the discount store into the epitome of chic.

William "Bill" Bernbach, 1950s.

Cartoonists captured a reality in which most national lenders—department stores, automobile companies, and banks—extended credit mostly to white men. Even as the consumer economy expanded, the disparities of who could fully enjoy the fruits of the postwar American dream came into sharp relief. With regard to credit cards, national lenders rarely lent to women, people of color, or the poor because they were seen as high-risk customers. Ironically, although women were often viewed as the primary consumers in a household, married women found it difficult to get credit cards in their own names. Banks and department stores demanded credit histories to determine risk and used the lack of such to discriminate against women, immigrants, and all people of color. And even if a woman held an account at a department store, she would be dropped or required to change her account to her husband's name once she married.

By the 1960s, this practice so infuriated certain white middle-class feminists such as Bella Abzug and others that they called on the FTC (Federal Trade Commission) to study and challenge gender-based discrimination in lending. Their efforts resulted in the 1974 passage of the Equal Credit Opportunity Act (ECOA). Professional women felt the sting of discrimination and had the clout, money, and organizational support to change the system. Their fight for full consumer rights resulted in more transparent and less discriminatory lending practices across the board.

Racial discrimination in lending became the next target, and Congress passed an amendment to the ECOA in 1976, prohibiting discrimination based on race. Most banks quickly eliminated discriminatory lending practices, especially since they ultimately stood in the way of profits. For instance, the National Bank of North America ran an ad appealing to women in 1974 that claimed, "whether you're a Miss, Mrs. or Ms., we make loans to all creditworthy people" (Hyman 2011).

Growing reliance on computer networks to track credit also allowed national banks to replace loan officers, who sometimes imposed their own

"Your Diner's Club card came today," by Whitney Darrow Jr., circa 1962. Whitney Darrow Jr., a long-time cartoonist for *The New Yorker*, played with the common image of women as spenders of their husbands' money, but also captured the reality that most married women could only have credit in their husband's name.

prejudices, with computerized credit scoring. Some lenders used zip codes to redline certain demographics. Diners' Club and two providers of gas credit cards were sued in the 1970s. Most banks, however, stopped using zip codes as a determinant of creditworthiness in the late 1970s and 1980s.

CREATIVE REVOLUTION

In the 1950s, the advertising industry typified the conformist, "other-directed" corporate culture critiqued by sociologists, novelists, and industry insiders. In fact, there were more novels and social critiques of advertising written in the postwar period than in any other time in history, including the iconic 1955 novel by Sloan Wilson *The Man in the Gray Flannel Suit*.

The creative revolution of the 1950s and 1960s remade the advertising industry, from staid to hip. Ads were irreverent, ironic, self-referential, and sometimes difficult to decipher. Agencies hired more diverse staffs and formed teams of copywriters and artists who worked together and measured their success not only on sales but also on the originality of their work.

Advertisement for Orhbach's department store by Bill Bernbach, 1958.

WOMEN IN BUSINESS

In April 1953, *Fortune* magazine profiled Margaret Dalby Brown, "*one* of 19 million" women in the United States who worked outside the home. Hired in 1943 as a riveter in a war plant, Brown liked the money and stayed on after the war. She and her husband bought a house in Levittown, Long Island, joining the growing ranks of two-income families. *Fortune* asked: Why did she work? Typically women left the workforce when they married, discouraged by both societal norms and business policy. Brown worked, *Fortune* explained, because she "wants the things most Americans want: a home of her own, security and 'advantages' for the children, good food, and clothes, a good car, a television set, and money in the bank" (Hamil 1953).

Employment of married women grew and by 1960 about 30 percent of wives and mothers worked outside the home, attracting the attention of opinion makers, marketers, and business leaders. The growth in women's employment was concentrated in two sectors, the professions and the low-wage service sector. Middle-class women entering business, law, and medical schools increased dramatically after legal gains in the 1970s, going from 5 percent after World War II to almost 40 percent by the 1980s. But most white women still worked in lower-wage jobs as secretaries, stenographers, clerks, retail saleswomen, and waitresses, meaning that despite their increased numbers in the workplace, those workplaces remained segregated by race and gender.

In the 1950s, at the leading edge of these trends, the census figure of 19 million women in the workforce reverberated through the pages of magazines and newspapers, as well as through the corridors of marketing firms. The lives of working women differed substantially from the prevailing media image set by June Cleaver, the stay-at-home mother in the classic television show *Leave It to Beaver*. For instance, sociologist C. Wright Mills declared in his widely read examination of the new middle classes, *White Collar*, that "It is as a

secretary or clerk, a business woman or career girl, that the white-collar girl dominates our idea of the office. She *is* the office...." (Mills 1951). These working women had new needs as consumers; they worked a 45-hour week in the office or factory and juggled domestic responsibilities at home. The working woman of the 1950s and 1960s, as marketer Estelle Ellis noted, had more money than time.

For her part, Ellis, after graduating from Hunter College, embarked on a career in publishing and marketing in New York City, and apprenticed to then well-known editor Helen Valentine. After starting *Seventeen*, a new magazine for teenaged girls, Valentine, Ellis, and a small staff of women writers, editors, and marketers started a new venture, *Charm* magazine that bet on those 19 million

***Charm*, August 1950.** *Charm*, published beginning in the 1940s, relaunched as a magazine for working women in 1950. The inaugural issue of the magazine featured a well-dressed model, the image of the sophisticated office workers that the editorial staff wanted to speak to through its stories on fashion, literature, and work.

Charm **department store display, 1950s.** *Charm* magazine partnered with retailers to host promotional events that featured clothing for the office. Such marketing events brought women into the stores and encouraged them to subscribe to the magazine.

working women as a unique and growing market. The first issue of *Charm*, with its new staff, debuted in 1950, the cover graced by a well-groomed model, the personification of the women professionals the magazine sought to reach.

Although *Charm* spoke to a cohort of white working women, the magazine touted the broader spectrum of women in terms of occupation, family life, and reasons for working. Feature stories addressed the various opportunities and common challenges facing working women, along with articles and literary pieces written by the era's leading cultural critics, journalists, and fiction

authors. While it set high standards for content, *Charm* also became a conduit for reaching this new market, especially for clothing makers and retailers. Working women required different wardrobes for home, work, evening, and travel and this need for more clothing made them a lucrative market. Ellis commissioned market surveys of women who worked and used the information about their buying habits, coupled with census data on their increasing numbers, to define a distinct market. Dubbing the growing ranks of middle-class workers, "busy businesswomen," Ellis advised retailers, in particular, on serving this group of powerful

who snags more stockings?

who wears out more shoes?

Who snags more stockings? Who wears out more shoes?, 1950s. Estelle Ellis Collection, Archives Center, National Museum of American History. Estelle Ellis, marketing director at *Charm* magazine, created this brochure about data on working women consumers. She argued that working women bought more clothing than women who stayed home, and, thus, offered a lucrative market for potential advertisers. She used the brochure to sell advertising space in the fledgling magazine.

consumers. One of her most persuasive arguments, that these working women needed more time to shop, convinced some department store owners to keep their stores open later at night.

Women fashion designers, businesswomen in their own right, created a new American look in the 1950s, one that tailored sportswear and business attire to a new generation of women. Claire McCardell (1905–1958), a leader in making ready-to-wear fashion for American women, had established herself as a designer in the 1930s and 1940s. A graduate of Parsons, she spent time studying in Paris, and apprenticed with Hattie Carnegie in the United States, but eventually rejected high fashion for an American style that emphasized practicality and comfort. She explained, "I like comfort in the rain, in the sun, comfort for active sports, comfort for sitting still and looking pretty. Clothes should be useful" (*New York Times*, 1958). A pioneer in many respects, she reshaped women's fashion, including developing the idea of separates. Separate skirts, pants, and tops gave women versatile, and economical, options; three or four separate pieces could be combined into a number of different outfits. McCardell's designs resonated with American women and she became one of the most successful designers of the 1950s, eventually creating her own label and winning numerous awards. As evidence of her success in business, *Time* featured her on its cover in May 1955.

Women who did not go out to work in an office, the service industry, or a factory sometimes engaged in profit-making enterprises at home. Some turned their living rooms into showrooms and earned money by selling new products such as Tupperware to their friends and neighbors. Housewives had a long history of making money from home, from selling eggs to taking in boarders, but in the postwar period industries that adopted home parties as a sales technique took off.

Although she didn't invent the home party, Brownie Wise (1913–1992) developed the form into a profit-making model that motivated women to tap into female networks and develop their inner pitchwoman. A wartime secretary at Bendix Corporation in the 1940s, Wise later sold goods

Claire McCardell suit, 1949–1950s. The washable knitwear, separate components, and neutral gray of this suit defined McCardell's take on ready-to-wear for women. Easy to launder, the separates could be combined with other pieces, making them economical. The cut and other details provided style.

Claire McCardell, **by Boris Chaliapin, 1955, National Portrait Gallery, Smithsonian Institution.** *Time* magazine artist Boris Chaliapin painted the famous fashion designer Claire McCardell for the cover of the magazine in 1955. McCardell revolutionized women's wear in America, designing practical but elegant clothing for a wide range of consumers.

for Stanley Home Products where she gained recognition as a top seller of Poly-T, or Tupperware. Earl Tupper, inventor of the plastic ware, hired Wise in 1951 to develop the company's home selling strategy, the Tupperware home party. Tupper had an innovative product but a serious sales problem. When he first introduced the sealable plastic containers, they didn't sell because buyers remained skeptical about plastic, which tended to

Postcard for Tupperware home party, 1957. Tupperware provided women with standardized tools for selling, such as postcards that invited potential customers to home parties, and that made buying and selling fun! Home parties were just that, parties, often with food, games, as well as demonstrations of the product.

Brownie Wise demonstrating Tupperware in Hawaii, 1950s. Brownie Wise, the dynamic personality behind Tupperware's home party system, made Tupper's plastic containers a household name in the 1950s. Her incentive programs rewarded top sellers and taught women how to realize their dreams through selling.

be brittle, and they didn't understand the seal. Americans housewives preferred the traditional glass containers that they knew how to use. However, Wise believed that home demonstrations could show the innovative, resealable container to best advantage and sell the vast array of bowls, cups, containers, and molds. Once convinced that home parties led to greater sales, Wise rose to vice president in charge of home sales.

In her new position, Wise created motivational techniques that worked especially well for stay-at-home wives and mothers. She made Tupperware a household name in the 1950s by creating a culture of home sales that rewarded top sellers with prizes and recognition as well as money. An annual Jubilee, hosted at company headquarters in Florida, recognized the most productive dealers and distributors, spurring them on through games and prizes, big and small, to do even more in the following year.

No matter where they worked, most women made less money than their male counterparts. Income inequality was the product of a complex mix of segregation of the workforce into secretarial and service jobs, and various barriers to advancement in the professions that kept women from achieving the highest-paid positions. Women active in labor unionism in the 1950s and vocal in the Kennedy administration of the 1960s pressed the issue of equal pay, which resulted in the passage of the Equal Pay Act (EPA) of 1963. The EPA mandated equal pay for equal work for women and men in the same jobs. Before the passage of the 1963 act, almost two-thirds of women worked in "clerical, service or sales" positions, only about 15 percent were managers, and all women earned roughly .59 cents for every dollar paid to men, although this gap widened for women of color (National Equal Pay Task Force 2013, 6). Feminists in the 1970s worked to eliminate barriers to higher education for women and the glass ceiling at professional jobs. By the 1980s, women made gains in pay and more women owned their own businesses, but wages remained below those of men in similar jobs.

RUTH HANDLER
1916–2002

Observing her daughter Barbara play with paper dolls, Ruth Handler recognized that her pretending focused on the dolls' future. Ruth was co-owner of the Mattel Toy company with her husband, Elliot Handler, and she convinced him to develop a teenage fashion doll. In 1959, the Barbie doll, named after her daughter, debuted at the New York City toy fair. With Ruth Handler's savvy merchandising the Barbie doll became the best-selling toy ever produced. The doll's shapeliness came under heavy criticism and informed the conflicting messages confronting young women at the time; Handler averred that the doll allowed girls to act out their future career dreams.

Ruth Handler, executive of Mattel Toy company, with collection of Barbie dolls, 1961.

MURIEL SIEBERT
1928–2013

"Part of my career goal was 'Where can I go where there is no unequal pay situation?' That's why I decided to buy a seat on the Stock Exchange and work for myself" (Kaputa 2009).

In 1967, Muriel Siebert took her seat on the New York Stock Exchange (NYSE), the only woman among 1,365 men. She faced a tough work environment: the NYSE charged her more for her seat than it did her male counterparts, and some banks refused to loan her the money; and the seventh-floor lunchroom of the Exchange did not have a ladies' room until 1987. Known to many in the industry as "the first woman of finance," she fought for many of her firsts.

Muriel Siebert, 1977.

Tupperware wish fairy laughing with women in audience, 1950.
Brownie Wise wanted saleswomen to feel special, and she created prizes and rewards along with memorable events to recognize top sellers. The annual Jubilee, a gathering of salespeople at Tupperware's Florida headquarters, used fun themes and games, including a wish fairy who gave out spectacular prizes to honor leading sellers.

BUYING POWER

Three years after the passage of the Civil Rights Act, on August 16, 1967, Dr. Martin Luther King, Jr., addressed the 11th Annual Meeting of the Southern Christian Leadership Conference (SCLC) in Atlanta and asked, "Where do we go from here?" While African Americans had made strides towards civil rights, economic rights still lay ahead. In answer to his own question, King praised the efforts of Operation Breadbasket, an SCLC initiative that challenged economic inequality in cities through boycotts. Operation Breadbasket, led by Jesse Jackson in Chicago after 1966, had made significant strides in pushing white-owned companies to provide jobs, advertise in African American newspapers, and make deposits at banks that loaned to African Americans. Above all, the SCLC and Operation Breadbasket, like many civil rights organizations, promoted a greater respect for African Americans as both workers and consumers. King reminded the crowd of Operation Breadbasket's slogan: "If you respect my dollar, you must respect my person" (King, Jr. 1967).

Although there was a growing awareness of African Americans as a market force during the great migration in the early twentieth century, and even a handful of market surveys on black consumer preferences, it wasn't until World War II that national businesses began to take seriously the buying power of African Americans. Until then and even through the 1950s, most businesses ignored, or worse, stereotyped African Americans in advertising, and subjected them to second-class treatment in stores. The SCLC understood that getting white businesses to advertise in black media outlets required a financial commitment to the African American community; it was also a sign of inclusion and respect. In addition, using accurate and respectful portrayals of African American consumers in national advertising made black consumers visible in ways they hadn't before been.

John H. Johnson (1918–2005), publisher of the pioneering *Ebony* magazine and later *Jet*, worked hard in the previous decade to convince national companies to advertise in his publications. When Johnson launched *Ebony* in 1945, he promoted circulation numbers and surveys documenting the buying power of his mostly middle-class readership. Most magazines relied heavily on advertising to fund publication costs, but few black publications could count on national advertising from white-owned companies. While General Motors, Coca-Cola, and others might have black customers, the companies rarely acknowledged them in advertising (neither advertising in black publications nor incorporating African Americans into advertisements for a general market) for three reasons: the companies believed that African Americans had low incomes and insignificant buying power; thought that African Americans would buy regardless of whether they saw advertising or not, and feared that by advertising to blacks they were degrading their brands in the eyes of what they considered to be more valuable white consumer markets.

Advertising agencies began to see the value of a growing middle-class African American market during World War II, when the Equal Employment Opportunity Act (1942) and military service and federal and manufacturing jobs increased African American incomes. Large advertising agencies created "special market" divisions and hired African Americans to study the market and advise on campaigns directed at African American shoppers.

David J. Sullivan, an influential African American marketer who wrote for a number of advertising and sales journals, had strongly worded advice on what to do to end the stereotyping in advertising of black Americans as servants— poor, rural, and uneducated—and on how to woo the growing urban, middle-class, brand-loyal market. In 1943, Sullivan published a detailed list of "don'ts" for white advertisers in an issue of *Sales Management* that systematically dismantled a century's worth of stock, black stereotypes in advertising—from waiters and cooks named George and Mammy to the white-haired, poorly clad

Pepsi-Cola Co., January 1954. Pepsi-Cola created a special markets division to advertise to African Americans in the 1940s, hiring African American marketing pioneers Edward Boyd and later, H. Naylor Fitzhugh. In this photograph, marketers discussed a campaign for Washington, D.C.

sharecropper known as Uncle Mose. "Do not picture the 'Uncle Mose' type," advised Sullivan. "The U.S. Chamber of Commerce says, and facts prove, that Negroes spend more money for clothes per capita than do white people in New York and other large cities" (Weems 1998).

Black-owned talent agencies hired live models for a new wave of advertisements that would replace the caricature portrayals of old with photographs of respectable middle-class women and men. By the 1950s, marketing trade journals and business publications including *Fortune* ran dozens of articles on the black consumer market. Most relied

on statistical data, comparing the rising incomes of African Americans relative to whites (about 50 percent of white income in the 1950s), and information on African American purchasing habits and preference for certain brands. Books with titles similar to William K. Bell's *Fifteen Million Negroes and Fifteen Billion Dollars*, published in 1956, used the profit potential of this market as a carrot for manufacturers, retailers, and advertising agencies to recognize and respect this group of consumers.

Advisers told retailers and manufacturers that radio was one of the best places to reach the African

HOW DO YOU
sell
TO URBAN NEGRO
CONSUMERS?

Brochure, "How do you sell to urban negro consumers?" Courtesy of *Ebony* magazine, circa 1960. *Ebony* magazine provided a window on the black middle and professional classes for national producers, retailers, and advertisers, and coached corporate America on how to reach these markets through brochures like this one.

AFRICAN AMERICANS AND ADVERTISING

African Americans began working in "special market" divisions of advertising agencies in the 1940s, advising businesses on how to reach the black middle class. By the 1970s, agencies run by African Americans and Latinos served regional and ethnic markets. These pioneers changed the look and feel of advertising and did groundbreaking research on these underrepresented populations.

Caroline R. Jones began her career in advertising at J. Walter Thompson, moved to one of the first black-owned agencies, Zebra, in the late 1960s, and eventually opened her own firm. Throughout her career, she urged clients to see African Americans as a varied market. She achieved many firsts in her lifetime and was often asked to describe what it was like at the top. Her answer: "It's cold but I can feel the sunshine."

Portrait of Caroline R. Jones at work, circa 1975.

The Consumer Era 1940s–1970s

147

American market. While *Ebony* achieved an unprecedented circulation rate by the 1960s, radio could reach even more potential customers. Radio was also a better deal for most advertisers than either magazines or the new mass medium of television. Both of these charged high rates for advertising, not all African Americans owned a television, and manufacturers were still shy of alienating white audiences with integrated commercials. Long before cable television enabled niche marketing, radio could narrowcast to local markets of African American listeners. Although the first black radio station started in 1929 in Chicago, black-oriented stations proliferated in the 1940s and 1950s, with the increase of more disposable income. Advertisers rightly identified these stations as ripe for advertising and as sources for gathering data on African Americans.

Starting in the late 1940s, *Sponsor*, the principal trade journal for broadcast advertising, ran a multiyear series of features on black-oriented radio, or "Negro-appeal" radio, so advertisers could better understand and reach what it called "The Forgotten 15,000,000." Using data from stations in the eastern half of the country, the research projected that radio ownership in certain large cities (New York, Washington, D.C., Philadelphia, Charleston, Atlanta, and others) was somewhere between 75 and 90 percent, indicating that the majority of black households owned radios. And, as data from at least one city showed, African Americans spent twice as much time listening to the radio as whites did. This should not be surprising as many public entertainment venues at the time, movie theaters for one, were still segregated by race. Radio could provide entertainment that was both free of potential embarrassment and inexpensive.

By 1955, there were 600 black-oriented stations in the United States, a national network, the National Negro Network, Inc. (NNN), and several regional networks. These stations played a mix of popular music and often featured gospel performers; they provided news and information of specific interest to black communities, including news and political opinion pieces that spoke to the concerns of their listeners. In addition, these largely white-owned stations relied on black deejays to be the face and voice of the station.

Black disc jockeys came into their own in the 1950s, providing a compelling mix of entertainment—new music, public service announcements, commentary, and salesmanship. Deejays played the latest music, and in many ways determined what groups made it onto the air. They helped define a generation of teenage listeners and consumers of music, and became local celebrities. Deejays also regularly promoted not just national products but local stores and retailers that catered to black customers.

Charles W. "Hoppy" Adams, was one such celebrity deejay. He worked at WANN in Annapolis, Maryland, one of the groundbreaking black-oriented stations that blossomed after World War II. Morris Blum, a white entrepreneur, war veteran, and staunch antisegregationist had founded WANN in 1948 to reach the black community and hired African Americans as deejays and managers at the station. Adams started at the station in 1952. A larger-than-life personality, he promoted black musicians, hosted WANN events at the then-segregated Carr's Beach outside of Annapolis, and broadcast from stores in the region to promote retailers and products to the black community. Adams worked his way up from deejay to executive vice president, later becoming a member of the governing board. While he was the voice of WANN, Adams embodied the complex threads of black-oriented radio with his respect for the community, its commerce, its music, and its burgeoning teen culture, all of which defined the programming at WANN.

Of radio stations serving Negroes exclusively . . .

WANN *serves the largest Negro market in America-outside of New York!*

If you're interested in the Negro market, here's a fact you might find interesting: According to the latest U. S. Bureau of Census figures, more Negroes live within the WANN coverage area than live within the coverage area of any other radio station in the United States serving the Negro market exclusively, outside of New York!

And WANN's impact in this market? We'd like to show you some astounding figures. Mail pulls! Surveys! Tremendous tangible responses to promotions and to commercials! Proof, beyond all question, of the overwhelming popularity of WANN in this area. It will amaze you!

Remember, too, WANN is the only medium selling to this Negro market exclusively. There IS NO OTHER MEDIUM serving this BALTIMORE-WASHINGTON-MD. EASTERN SHORE area this way, an area which embraces more than 600,000 Negroes spending more than $250,000,000 a year!

THE WANN Story makes quite a story. We'd like to show it to you. Call or write us any time!

1190 on the Dial

Annapolis Broadcasting Corp.

P. O. Box 749 ● Annapolis, Md.

Phone COlonial 3-2500

when you BUY ON
WANN
YOU BUY IN!

Write For Free Brochure!

$250,000,000 Negro Market Coming Up."
Yours for the asking on your letterhead!

WANN advertisement, circa 1950. WANN. Radio stations, like magazines, ran on advertising, and WANN worked to convince national companies to advertise over the airwaves. It was not an easy sell, and the fliers they sent out emphasized the spending power of the African American market that was worth $250,000,000 a year.

above, top: **WANN deejay "Hoppy" Adams broadcasting from a storefront window, Annapolis, MD, 1953.** WANN deejays connected their audiences to retailers and promoted consumption of not just music, but other goods and services. Adams conducted remote broadcasts from stores, which drew crowds of teen consumers and proved that WANN could be an important conduit for reaching the African American market.

above, bottom: **WANN remote broadcast set, late 1950s.** WANN deejays often broadcast from remote locations, including stores, connecting listeners not only to music but also to retailers.

As the Consumer Era began, the nation emerged from the longest depression in its history and World War II. Americans had renewed faith that capitalism, technology, and business would bring them economic prosperity. Wartime spending had put manufacturing and business back on their feet, and now these sectors turned to improving infrastructure and creating new products. Production boomed and consumerism shaped marketplaces, which spread from downtown business districts to new suburban shopping centers. Innovations in technology, expansion of white-collar jobs, more credit, and new groups of consumers fueled prosperity. With Europe and other competing nations in ruins, American business dominated the world's markets with a steady stream of new technologies and goods.

Americans had more money to spend and were eager to buy houses, cars, televisions, refrigerators, and hundreds of other products. Federal Housing Authority (FHA) loans and the GI Bill fueled prosperity by subsidizing the cost of housing. To realize the dream of middle-class life, many people turned to new forms of consumer credit and established two-wage earner households.

Between 1950 and the 1970s, more married women entered the workforce. A highpoint of unionized labor produced a blue-collar middle class that earned enough to buy new goods, take vacations, save for retirement, and send kids to college. Although fewer Americans were farmers, improvements in agriculture fed not only their fellow citizens, but also millions in other countries, leading many to think of the United States as a breadbasket of the world. While prosperity did not reach everyone, and pockets of economic insecurity and inequality remained, most Americans benefited from the economic expansion.

As years passed, growth was tempered by war, rising oil prices, and cultural anxiety. The Cold War turned hot in Korea and Vietnam. The costs were high, in both human lives and national resources, and debates about America's place in the world colored political rhetoric and popular culture alike. The Civil Rights Movement and the War on Poverty brought images of inequality into living rooms across the country and raised issues about the shortcomings of prosperity and the common good. Some younger Americans sought alternative lifestyles that questioned traditional definitions of success.

The economic engine slowed in the 1970s, leading Americans to engage in a more general questioning of the promises of consumer capitalism. An energy crisis in the early 1970s lengthened into an ongoing series of shortages and increasing prices for gas and a host of goods. At the same time, productivity waned, wages flattened, and "stagflation"—a combination of economic stagnation and inflation—made this decade a sharp contrast to the previous two. Economists used a "misery index," which combined the unemployment and inflation rates, to measure the impact of these economic changes on consumers. Many believed the problem was less a national malaise than mismanagement. President Jimmy Carter lost his reelection bid to Ronald Reagan, who would serve two terms and lead a conservative revolution on the promise of a new "morning in America."

In the 1980s, companies globalized and became multinationals, deregulated, and found vastly increased efficiencies through application of digital technology and networks. These changes paved the way to the new realm of online commerce. Consumers had more choices for low-cost goods and access to information from around the world. But many well-paying manufacturing jobs were off-shored or automated, and new jobs in the retail and service sectors were often low-wage and nonunion. Of particular importance in shaping the future was the growth of the financial sector. Entrepreneurs developed a wide range of new products to raise capital and move it across the globe at lightning speed. They, and the multinationals they served, would reshape the landscape of business into the twenty-first century.

Business and Consumers: What Makes for a Great Relationship?

SHEILA BAIR

While head of the Federal Deposit Insurance Corporation from 2006 through 2011, Sheila Bair worked through the Great Recession to stave off an even bigger crisis and to stabilize the economy. Frequently locking horns with top federal regulators whom she thought were being too easy on Wall Street, Bair has been keen to look out for consumers. Born and raised in Kansas, Bair began her career in Washington, D.C. as an attorney with former Senator Bob Dole. She's also worked on Wall Street and in academia, giving her a varied perspective on the checks and balances in our system, and those ever-present economic cycles.

How do you describe the complex connections between American corporations and the consumers they purport to serve? They are not homogenous. Some are healthy and some are not. Some are characterized by deceit and trickery; others, by abuse or co-dependence. The most successful ones seem to be marked by honesty, openness, and a long-term commitment to making the relationship work.

Take these three iconic American brands: Ford, Disney, and Apple.

Each began with a singular, simple idea for a product that met—or anticipated—deep-seated needs or wants of the American consumer. They were founded by visionaries looking for meaningful affairs with their customers, not frivolous dalliances. They wooed consumers with innovations that promised lasting benefits. In doing so, they built successful corporate empires that stood the test of time.

By finding a way to mass-produce an affordable horseless carriage, Ford revolutionized consumer mobility and freedom, connecting the rural with the urban, creating a new "suburban," and fostering a multitude of related industries—gas stations, motels, roadside restaurants—all catering to the needs of the traveling woman and man.

As the automobile and other innovations gave American workers more leisure time, the Disney Company helped them fill it with new forms of entertainment, beginning with a moving picture that synchronized sound with beautifully crafted animations. Disney's irresistible Mickey morphed into a media kingdom, providing quality experiences that could be shared by the entire family and giving multiple generations of Americans common cultural references in big-eared mice, lisping ducks, sleeping beauties, and boys who never grew up.

Finally, Apple tapped into our inner child by developing a new kind of computer that was more toy than machine, as beautiful to the eye as it was friendly to the user. It displaced the multi-cabled, plastic box monsters and their 200-paged manuals that had previously confined cyberspace to the nerdy elite. More than any other computer company, Apple democratized technology and fundamentally changed the way individual consumers communicate and interact with the world.

Each company started with a simple vision to capture not just American consumers' wallets, but their hearts and souls. Each grew into lasting corporate empires that offered a plethora of products that hewed to that vision. Over the years, if they stumbled, it was because executives put deal-making and dividends over consumer interests, as did Apple in the late '80s and early '90s, or they lost touch with their customers, as happened at Ford. Steve Jobs' triumphant return to Apple in 1996 showed that the best way to serve shareholders is to ignore them and focus on customers. With that strategy, he turned Apple into the most valuable company in the world. Similarly, Alan Mulally turned Ford around by refocusing the company on its customers, creating "One Ford" and disposing of the exotic foreign brands the company had purchased over the years. Mulally has famously said that he identified the company's core problem on his first day of work when he drove into the executive garage and saw it filled with Jaguars and Aston Martins.

If the best companies in America have wanted lifelong bonds with their customers, the worst ones have had something altogether different on their minds. Our economic history is replete with corporate lotharios who take advantage of the financially naïve and innocent—consumers eager to believe in things too good to be true. They may take from Peter to pay Paul, à la Charlie Ponzi or Bernie Madoff. They may pervert a traditionally safe product—like a 30-year fixed-rate mortgage—into something quite explosive, like Angelo Mozilo's "hybrid fixed" home loans. Or they may start as legitimate and evolve into something quite different, like Ken Lay's Enron, which went from respected energy company to ethical cesspool, lying to investors, manipulating electricity prices, and gleefully deriding their consumer victims as "Grandma Millie."

Monopolists, on the other hand, are only interested in domineering relationships. They want to make profits not by keeping their customers happy, but by preventing other companies from vying for their affections. Sometimes they are undone by government. But usually, more nimble suitors find a way to slash through the anti-competitive thicket they have built and swoop away their unhappy customers. Standard Oil, AT&T, and Microsoft were all ultimately outdone as much by market dynamics as government lawsuits. Just as in the future, big cable may well be undone by satellite, Wi-Fi, and content companies, credit card monopolies by frictionless virtual currency, and title insurers by electronic value chains.

Even successful unions between corporations and consumers—while satisfying to both parties—can cause damage to others. After all, the Macbeths had a good relationship too, but look at the havoc their love wrought. By avoiding taxes, using foreign sweatshops, or disregarding environmental impact, corporations can reduce prices for their consumers (and increase shareholder profits), but society ends up covering the costs. Yet, for this, both consumers and their corporate partners bear responsibility. If consumers preferred corporations that paid taxes, protected the environment, and treated workers fairly, the market would not tolerate such abuses. At times consumers have used their purchasing power for the greater good, but more often than not, they look only for the lowest price.

And with declining wages and yawning income inequality, who can blame them for being careful with their dollars? The consumer product innovations of the early and mid-twentieth century created a virtuous cycle. As consumer demand for goods increased, so did the demand for labor to produce those goods, creating more jobs, more income, and yet more consumer demand. In the late twentieth century, however, that circle was broken. With the advent of globalization and technology, more goods could be produced with fewer U.S. jobs. Government responded by entering into its own dysfunctional liaison with consumers. It tried to keep consumers spending with housing bubbles and easy access to loans. Addicting consumers to credit created an unhealthy, and unsustainable, co-dependency.

It ended in tears with the 2008 financial crisis.

A half-decade later, the government is still trying to woo consumers with cheap credit to resuscitate our still-ailing economy. But consumers need and want a more stable relationship based on a strong job market and real wage growth. The virtuous circle of the twentieth century has been replaced with a negative feedback loop, as the shrinking purchasing power of un- or underemployed workers takes its toll. Even if government can't grasp this new reality, corporate leaders must—for their own survival. What good is it for corporations to keep driving down labor costs? Unless they make yachts or private planes, who is going to buy their goods?

In the past, success in corporate and consumer relations has been defined primarily by quality, affordable products. But change is a given. One partner must adapt to the evolving needs of the other, and in the future, it won't be enough for corporations to simply produce wonderful things for consumers to buy. Corporate innovators will need to think more holistically about who their customers are. Yes, they are purchasers of their products, but they are also taxpayers, families whose health is impacted by the quality of the environment, and workers who must earn a decent living to buy their goods. Technological advances can be a positive, not a negative, for workers, if it means shorter work weeks, safer workplaces, and better compensation for their skills. Henry Ford recognized this over a century ago when he raised worker wages and cut their hours, not only to improve productivity, but also to give his employees more purchasing power to buy Ford cars.

Corporate leaders need to fundamentally redefine their relationship with consumers. We need leaders who understand that their company's long-term success will be based on a tax system that more fairly distributes its burdens between workers and corporate owners; in an environment that fosters consumer health and ecologically sustainable growth; and, most of all, on compensation structures that revive our struggling middle class. Better pay, better training, and better opportunities for work-sharing and profit-sharing would be good places to start.

Informed consumers need to shower their affections on companies led by such visionaries. That would truly be a marriage made in heaven and one destined to last.

Who's Minding our Business?

SALLY GREENBERG

Sally Greenberg, Executive Director of the National Consumers League, knows better than most how the tension between businesses and the consumer is as old as our nation itself and will not disappear any time soon. Founded in 1899, the NCL is the nation's oldest consumer organization. It advocates for consumers in areas like fraud, food safety, medicines, and privacy. Greenberg has worked as an activist at different organizations for decades in Washington and Boston, including recently at Consumers Union where she worked on product liability and auto and product safety issues.

Business enterprise in the United States consists largely of companies producing goods and services for consumer consumption. Consumers want products that are reasonably priced, well made, and useful. Businesses may indeed be interested in producing a quality product, but they also seek to maximize profits. These goals can sometimes align, but many times they diverge. Profit motives and concerns about maximizing investor returns can result in the market failing to protect consumers from, for example, adulterated drugs or food, deceptive marketing practices, scams and rip-offs, and shoddy workmanship.

Consumer advocacy and the consumer protection movement helped to keep the marketplace in balance, serving as an essential check of power on business interests. Federal statutes passed at the turn of the twentieth century created the first consumer protection laws and helped spawn a vibrant movement that today is represented by a broad-based group of advocates that includes individuals and organizations. There will always exist a tension between the profit-driven motives of business and the safe and effective consumption of goods. Advocates seek a balance between these competing interests.

Consumer advocates work with local, state, and federal legislators, regulators, judges, and the courts to enforce existing safety standards and create new regulations to protect consumers. Advocates want to equip consumers with the tools and information they need to recognize and compare the quality of goods and services.

The National Consumers League, the oldest consumer and worker advocacy organization, founded in 1899 and led by the indomitable Florence Kelley, was at the forefront of those efforts. The League fought for passage of the Pure Food and Drugs Act of 1906 and the Federal Meat Inspection Act of 1906, both of which were signed into law on the same day by President Theodore Roosevelt.

The Pure Food and Drugs Act defined "adulteration" and "misbranding" and made illegal the production and shipment of misbranded or adulterated food. Producers of common cold remedies proudly advertised high dosages of caffeine, alcohol, cocaine, heroin, opium, and other drugs. Mrs. Winslow's Soothing Syrup, for example, intended for infants and children, claimed to "relieve the poor little sufferer immediately." The mystery cure? Morphine. Six years after passage of the Pure Food and Drugs Act, Congress passed the Sherley Amendment, which made it illegal for medicines to be labeled with false therapeutic claims. The Soothing Syrup, which the American Medical Association called a "baby killer," was finally outlawed in America in 1912. Laudanum, a drug that claimed to heal ailments ranging from menstrual cramps to insomnia, advertised on its packaging 40 percent alcohol and 47 grains opium. Laudanum still exists today, but the ingredients have changed, it is now only available with a prescription, and it is tightly regulated by the FDA.

Early efforts to protect consumers landed advocates in the crosshairs of business. In 1912, *The American Food Journal* called the National Consumers League a "fraudulent American organization, which represents no consumers." The Progressive Era was a turning point for consumer advocates who persuaded policymakers that consumer protection laws must hold businesses accountable for products and services.

In 1936, Colston Warne, with the help of a group of professors, labor leaders, journalists, and engineers, founded Consumers Union, which published *Consumer Reports*, a magazine to test the safety and quality of everyday products. Warne became the most influential consumer leader in the country. He actively promoted the spread of grassroots consumer organizations and found ways to support them. *Consumer Reports*'s testing changed the face of consumer protection. The magazine, which quickly gained hundreds of thousands of subscribers, refused paid ads, freeing it from bias and undue influence.

Consumer Reports tested seatbelts in the 1950s and found that two-thirds of them failed; tested cigarettes in the 1950s, separating tobacco manufacturer hype from reality and revealing serious health risks from smoking, findings later used by the U.S. Surgeon General; and found strontium-90 in samples of milk, fallout from nuclear weapons testing. In the 1970s, under attorney Rhoda Karpatkin's leadership, *Consumer Reports* discovered early microwave ovens to be leaking radiation. More recently, in the 1990s, the magazine tested SUVs and found certain models to be prone to rollover.

It's hard to imagine the modern-day consumer movement without the pioneering work of Ralph Nader, who gained national attention in 1965 with his exposé of the auto industry, *Unsafe at Any Speed*. The book indicted General Motors for the Corvair's suspension design that made handling extremely dangerous. In an effort to cut costs, GM eliminated a front stabilizer bar which caused the wheels to tuck under during turns. General Motors hired private detectives to tap Nader's phone and employed "call girls" in an effort to discredit him. Nader couldn't be compromised, and the campaign was exposed in Senate hearings.

Nader exemplified a classic David fighting the corporate Goliath. Idealistic college and law grads flocked to Washington to join the consumer movement he gave life to—in 1970, 30,000 students, including a third of Harvard Law School graduates, applied for 200 jobs posted, working for reforms in taxes, auto safety, insurance, pension, open government, and many more issues.

Consumer advocacy was instrumental in the establishment in 1970 of the National Highway Traffic Safety Administration (NHTSA) to regulate auto and highway safety. From 1979 to 2005, highway deaths decreased nearly 15 percent due to safety technologies like air bags, more crashworthy vehicles, and stability controls. In 1973, the Consumer Product Safety Commission (CPSC) was established, and thirty years later, the agency

estimated a 30 percent reduction in deaths and injuries from consumer products, including a 92 percent decrease in child poisonings and crib deaths.

Some might argue that with federal consumer protection agencies established, the work of consumer advocacy is done. Nothing could be further from the truth. Consumer advocacy is as necessary today as it was 100 years ago. Many issues remain and others are more complicated and have global reach. For example, as more private information gets stored online—annual global data production is expected to hit 35 zettabytes (or 35 trillion gigabytes) by 2020—hackers are increasingly successful in gaining access to sensitive personal and financial information. Consumers are also being forced to give up the right to sue companies for wrongdoing when, as a condition of buying a product or service, they must sign a contract with a mandatory arbitration clause buried in the fine print. These clauses are ubiquitous and are found in cable TV, cell phone, mobile home, medical service, and countless other contracts. They serve to protect corporate wrongdoers from accountability. Low-income consumers face vast exploitation with financial products like pay-day loans or car title loans that carry crippling interest rates. These are some of the many challenges that lie ahead for consumer advocates.

Capitalism by definition means that business will seek to maximize profits, often at the cost of consumer safety and protection. Consumer advocacy thus plays a critical role in keeping a check on business, helping to ensure that our products are safe, that interest rates are kept in check, that products sold to consumers are what they say they are, and that corporations can be held accountable in our courts.

There will always exist a tension between capitalism and consumer protection. Consumer advocates will always need to serve as watchdogs, insisting on enforcement of existing regulations, closing loopholes in current laws, and writing new legislation for emerging problems.

Intel clean room, Hillsboro, Oregon, 2014. A semiconductor "fab" (a clean room complex where computer chips are manufactured) is one of the world's most expensive factories to build. Here, computer chips, barely a few atoms thick, are fabricated using robotic systems in super-clean air.

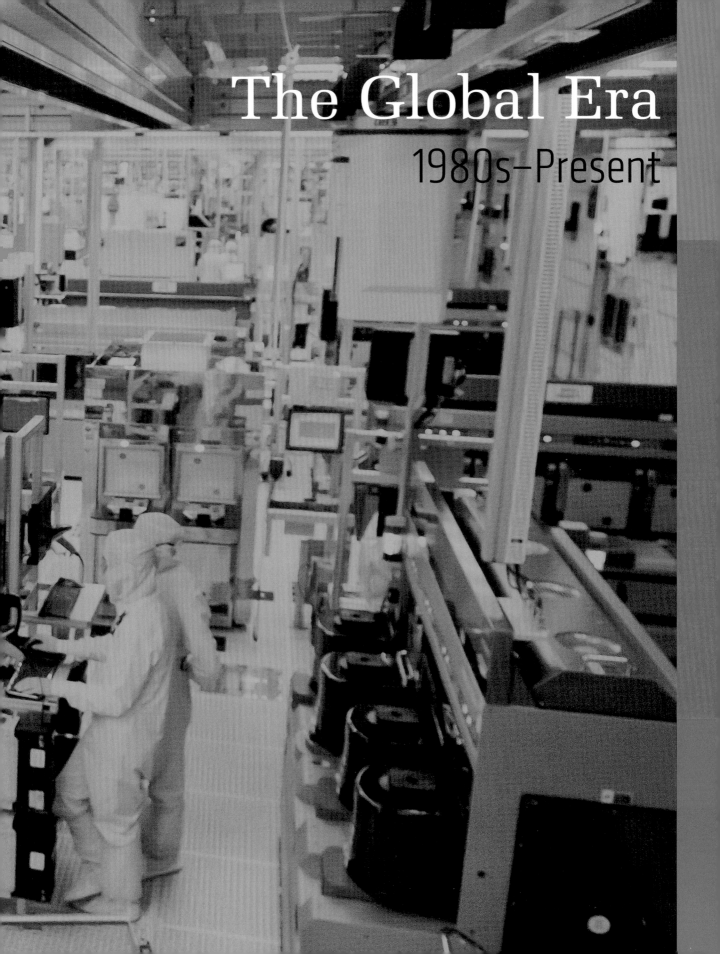

The Global Era

1980s–Present

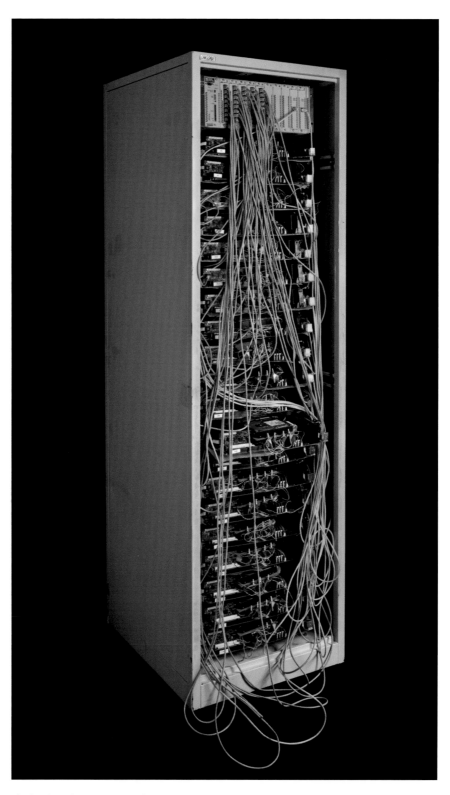

Google server, 1999. Stanford University graduate students Larry Page and Sergey Brin launched Google from a garage in Menlo Park, California. They cobbled together thirty racks of servers, including this one, from cheap computer parts to power their Internet search engine. They used venture capital to build their business.

THE GLOBAL ERA surrounds us, which makes it difficult to see in historical perspective. Yet even through the haze of current events, it is clear that some patterns of earlier eras of business history have persisted. Americans remained strongly committed to both capitalism and the common good. U.S. businesses continued to prosper throughout most of the period, but society again began to worry about the division of the profits of success and the effects of competition. Despite challenges, immigrants still regarded the nation as a land of opportunity where they could achieve material prosperity and social mobility. Innovation continued to drive economic expansion.

In these recent times, consumers and producers alike had to adapt to rapid change driven by computers, large amounts of data, and smart devices; by 2015 life without access to the World Wide Web and social media seemed for most a distant memory of a bygone era. Increased global interdependence created efficiencies and opportunities for some, but erased safeguards and employment for others. The decline in size of the manufacturing sector and the rise of the retail, service, and financial sectors had huge effects on American life. Immigration increased rapidly and again became a towering force in America—changing the face of the nation and providing a diverse pool of entrepreneurs and workers.

Another dominant trend was reduced government regulation. Taking office in 1981, President Ronald Reagan set the tone for the era with the implementation of supply-side economic reforms. Reagan reduced what he saw as barriers to economic growth—taxes and government regulation—in an effort to create a free market economy not seen in the United States since before the New Deal. The United States became a nation of fewer limits, but also fewer guarantees.

above, top right: **Union buttons.** In his first year in office President Ronald Reagan put labor on notice by firing 11,000 members of the Air Traffic Controllers Organization (PATCO) when they went on strike. The power of unions varied. Some grew, while others like the Amalgamated Clothing and Textile Workers Union (ACTWU) declined.

above, bottom right: **Milton Friedman's briefcase, about 2004.** President Ronald Reagan based much of his economic policy on the arguments of Nobel Prize–winning economist Milton Friedman. A staunch advocate for small government, less regulation, and free markets, Friedman (like Adam Smith 200 years earlier) believed self-interest and competition lead to economic prosperity.

GLOBALIZATION

Globalization—the expansion of global trade by multinational companies—created both opportunities and problems. For consumers and producers globalization lowered prices and improved the variety of goods, altered systems of production, and affected the availability of jobs. Computer networks improved the ability of businesses to manage global supply chains and interpret massive amounts of data. Globalization brought consumers gains, but at the expense of local business. Global trade created economic interdependence among nations and improved opportunity for many, but challenged local cultures and customs. For many people around the nation and the world, life was thrown into turmoil as the balance between traditional (local) and international (efficient) seemed elusive.

Tiffany glass paperweight, 2001. In the late 1990s, globalism became a management fad. Some companies gave senior managers objects to remind them that they were a global enterprise. This damaged paperweight, found in the debris from the attacks on the World Trade Center on September 11, 2001, shows the results of globalization gone wrong.

Name brands played a visible and often symbolic role in the global story. Seeking to sell more products, American companies often sought foreign markets. Similarly, foreign companies sought to enter the gigantic U.S. market. Coca-Cola serves as a good example of globalization. From the early twentieth century onward Coca-Cola spent heavily on advertising and successfully exported its products, making it the world's most recognizable brand. Some people drank the soda to be Western and hip; others rejected it as a symbol of American culture and power, and many just liked how it tasted. In 2003, after the U.S. invasion of Iraq, a Turkish company, Ülker, introduced Cola Turka. It challenged Coca-Cola's 57 percent share of the Turkish drink market by building on anti-American sentiment and instilling Islamic "positive nationalism." Ironically, some of the new brand's initial television advertisements featured American actor, Chevy Chase, as a New Yorker who sprouts a beard and starts speaking Turkish after drinking a can of Cola Turka.

Committing a national economy to a global way of life can present complex problems. Just about every country is in favor of unrestricted exports, but most countries also seek to protect their domestic industries and workers. Tariffs, quarantine rules, and trade barriers abound. In 1944, as World War II drew to a close, leaders of the Allied nations came together at the Bretton Woods Conference in Bretton Woods, New Hampshire, seeking to diminish trade barriers, regulate the international monetary system, and create a new financial order. The General Agreement on Tariffs and Trade (GATT), the World Bank, and the International Monetary Fund (IMF) were founded by the Allied nations, but trade issues still abounded.

The desire for open markets and lowering economic nationalism was furthered in 1995 during the GATT Uruguay Round of trade negotiations with the creation of the World Trade Organization (WTO). The international organization took the place of GATT, dealing with the regulation of trade among participating countries and providing a

CABLE TELEVISION AND MTV

As cable television emerged in the early 1980s, the innovative business allowed for a new realization of an older concept: narrowcasting. Cable channels, such as MTV (standing for music television) launched in 1981. It enabled advertisers to target specific youth markets. MTV competed in the fierce media market through innovative, frenetic advertising and must-see videos that defined their youthful audience and put off older viewers, creating a narrow community of special-interest viewers. Although the music channel had difficulty securing sponsors at first, it ultimately achieved success; hip hosts introduced edgy music videos, promoted artists, sold music, and positioned brands in a seamless mix of performance, video art, and advertising.

above, top: **Cola Turka, 2012.** Ülker marketed their cola drink as "positive nationalism." Other examples of coca-colonization protest products include the 2002 launch of Mecca-Cola (which campaigns against "America's imperialism and Zionism") and Qibla Cola (which provides "an ethical alternative").

above, bottom: **Coca-Cola, 1999.** Coca-Cola is the most widely recognized brand in the world. In 2013, 63 percent of its $46.7 billion in sales came from over 200 countries around the world. To maintain its brand awareness the company spent $3.3 billion on advertising.

MTV advertisement, about 1990.

ADVERTISING IN THE AGE OF CONGLOMERATES

Advertising had a global footprint from the late nineteenth century on, but the size of agencies reached an unprecedented scale in the 1980s and 1990s. The business of advertising changed dramatically in the mid-1980s with what industry journalists called "the big bang"—the merger of several large agencies into one multinational holding company, Omnicon Group. A wave of mergers created five large holding companies. At the same time, however, small independents and regionals came of age, did cutting-edge creative work, and competed with the big firms. Advertising also inspired debates about the power of American capitalism to homogenize world cultures. Does the constant barrage of ads from a handful of global brands obliterate local culture and make us all the same? Or does advertising provide the materials from which culture is made?

Nike T-shirt, 1980s.

framework for negotiating and formalizing trade agreements, as well as a dispute resolution process aimed at enforcing participants' adherence to WTO agreements. But rules around international trade tend to be complicated and often operate without direct community input, making local citizens angry. Invisible to most Americans, negotiators worked with foreign governments and interested parties to create trade agreements and resolve disputes. The move towards global free trade and fair competition rather than supporting local traditions and desires has spawned numerous protests.

In the 1990s, arcane trade talks became the center of heated family discussion as North American Free Trade Agreement (NAFTA) negotiations and WTO protests made the evening television news. Trade negotiators, seeking to improve their country's position, often ended up in the middle of unexpected battles. Consumers and economists generally favored free trade because the lack of tariffs and quotas made imported goods cheaper and promoted efficient production. On the other hand, many politicians, seeking to maintain jobs and cultural traditions, liked protectionist policies. Workers' reactions varied depending upon whether they were affected.

Trade tension reached a public peak during the 1999 World Trade Organization conference held in Seattle, Washington. Violent demonstrations broke out in the streets as anarchists, labor, and social activists came to protest the actions of what they believed was a sinister and tyrannical organization. The efforts of the nongovernmental organizations (NGOs) were especially interesting as their actions provided insight into the range of impact of trade rules.

The Humane Society of the United States organized early and came to Seattle to argue that environmental law should come before trade law. Their turtle costumes became one of the icons of the gathering. A decade earlier, in 1989, the U.S. Congress, seeking to protect endangered sea turtles, had banned the import of shrimp from any country

Protest buttons, about 2000. The World Trade Organization (WTO), International Monetary Fund (IMF), and World Bank operate without local input. Invisible to most Americans, negotiators work directly with foreign governments and interested parties to create trade agreements and resolve disputes. A global rather than local vision has spawned numerous protests.

Turtle costume, 1999. Environmentalists' simple but showy costumes were widely photographed by the media. The Humane Society of the United States argued "The WTO may have been great for free trade, but as far as animals are concerned, the WTO is the single most destructive international organization ever formed" (The Humane Society of the United States 1999).

World Trade Organization protest in Seattle, November 30, 1999. Wearing sea turtle costumes made out of painted cardboard, activists from the Humane Society protested the WTO putting free trade before environmental laws.

that did not require turtle excluders on shrimp fishing nets. Mexico, Malaysia, India, and Pakistan objected, filing a WTO protest in 1996 claiming the cost of installing the new equipment was prohibitive and created an unfair trade barrier. The four countries felt that the United States should not be able to dictate their national environmental laws. The WTO agreed, and held that the U.S. ban was an illegal trade barrier. The demonstrators objected to free trade coming before the environment and animal protection.

Typically, most trade battles are over simpler desires to protect industries. Factory owners and farmers often seek protection when more efficient or lower-paid workers in foreign countries challenge their products. The manufacture of clothing is an excellent example. With little money needed to set up a garment sewing operation and the advent of fast, secure, and inexpensive

Mexican Coca-Cola, 2002. From 1997 to 2006, Mexico and the United States had a bitter battle over Mexico's sugar industry. Twice, Mexico tried to protect its industry by taxing U.S. high-fructose corn syrup used in Mexican soft drinks. The WTO sided with free trade rather than protecting local farmers.

below: **Trade barrier poster, 1999.** In 1989, the European Union banned the use of hormones in cattle (used to increase production and lower the cost of meat) because of health concerns. The United States challenged the ban and the WTO declared the hormone ban an unfair trade barrier.

containerized shipping, the once strong U.S. apparel industry came under siege from offshore producers in the 1980s. As U.S. retailers became more adept at managing large supply chains through scanning and telecommunications, the shift to offshore garment production accelerated. Consumers, attracted by low prices, bought more goods.

Interestingly, the move to outsourcing American apparel (and other manufacturing) actually was fostered by the U.S. government. In the 1950s the U.S. federal government sought to improve economic conditions in Puerto Rico (a commonwealth of the United States) by promoting a transition from an agrarian base to an industrial society. Washington bureaucrats created Operation Bootstrap (the Industrial Incentives Act of 1947), which encouraged American companies through tax incentives to move production facilities to Puerto Rico. Many companies soon discovered the advantages of cheap and docile offshore labor and moved production even further offshore. Outsourcing was exacerbated in the late 1970s and 1980s as many American family-run companies with close ties to their communities sold out. High-yield bonds created by financial pioneers like Michael Milken created pools of cash utilized by private equity firms like Kohlberg Kravis Roberts (KKR) to let owners sell their companies and retire (a system KKR called bootstrapping). Now publically traded companies run by professional managers and answering to the demands of investors instead of family ties, the companies often made their production more efficient by moving production offshore.

Some see access to the large U.S. market as an opportunity for developing countries. Others argue that it exploits both domestic and foreign workers. Free trade can make the wages of workers spiral down in order to compete with overseas labor. According to an activist group, the National Labor Committee, H.H. Cutler (a subcontractor to the large apparel producer VF Corporation) paid Haitian workers $0.06 per piece (about $0.40 per hour) for every $19.99 pair of *101 Dalmatians* pajamas sold in 1996.

GORDON AND CAROLE SEGAL
1939–Present

While on their honeymoon in St. Thomas, Virgin Islands, in 1961, Carole and Gordon Segal were impressed by a store that offered modern Scandinavian designs at affordable prices. Returning to Chicago, they soon decided to move away from their somewhat unexciting careers (Carole teaching and Gordon in real estate) and open a home goods store targeting people like themselves—young consumers with good taste but limited income. With no venture capital available, they took their wedding gift money (about $10,000), a small loan from Gordon's father, and opened a trendy store in Old Town in Chicago. Undercapitalized, they were forced to be innovative and limit expenditures as they set up the store in an old elevator factory. One creative cost-cutting choice was saving the shipping barrels and containers from their merchandise, reusing them as store display furniture. They were so pleased with this look they named their store Crate & Barrel.

Arzberg teapot, 1959. While dating, Carole helped Gordon select this teapot as a gift for his mom. The experience of shopping together helped established their collective taste and eventually the direction for their store—Crate & Barrel.

MICHAEL MILKEN
1946–Present

Revered by some and reviled by others, workaholic Michael Milken pioneered major financial innovations in the 1970s and 1980s, transforming access to capital and reforming corporate America. Running the high-yield bond department at the investment firm Drexel Burnham Lambert, Milken acted on economist Walter Braddock Hickman's underappreciated philosophies including his 1958 book, *Corporate Bond Quality and Investor Experience*. Milken believed that high-risk, noninvestment-grade, bonds were a better choice for a diversified portfolio than safer low-interest, investment-grade, bonds. His high yield (junk) bonds raised large amounts of capital for risky start-ups (like MCI Communications and Turner Broadcasting) and leveraged buyouts (like Beatrice Foods and RJR Nabisco). One of the highest-paid financiers in history, Milken earned more than $1.1 billion between 1983 and 1987.

Michael Milken on the corner of Broad and Wall streets, New York City, 1975.

Children's pajamas, 1996. These Walt Disney Company's *101 Dalmatians* pajamas were sewn by garment workers in Haiti being paid about $0.40 cents per hour. The design of the pajamas, along with the weaving, printing, and cutting of the fabric, was done outside of Haiti.

Activists and concerned citizens seeking to curb exploitive production copied tactics pioneered in the fight against racism in South Africa. Beginning in the late 1970s, many institutional and individual investors battled apartheid by divesting the stock they owned in companies doing business in South Africa. Today, several mutual fund companies offer "socially responsible" investment portfolios that do not include companies involved in sweatshop production and challenge corporate responsibility.

At the Walt Disney Company's 1997 annual meeting, Progressive Asset Management, Inc., brought to a vote a shareholder resolution governing suppliers' labor practices. Although the resolution did not pass, it received surprisingly strong shareholder support (39 million shares, or 8.3 percent). Subsequently, Disney pledged to issue and post a contractor code of conduct and authorize audits and inspections of all contract facilities. Disney garment contractor H.H. Cutler then pulled out of production in Haiti.

Imports to the United States are not always bad for American workers as new ideas and equipment from overseas can positively impact American industry. Management techniques in the auto industry provide a great example. In the 1980s, Japan challenged Detroit's dominance in the production of cars. Fearing Japanese gains in automobile sales, General Motors (GM) joined Toyota in a joint venture—New United Motor Manufacturing, Inc. (NUMMI). Working together with the Japanese in a Fremont, California, plant, GM executives learned and adopted Japanese management techniques like teamwork and kaizen (worker-led improvement). Importing cars took away American jobs, but importing the innovations of Japanese management style and adapting them to American manufacturing improved competitiveness.

Autoworker's hat, about 1988. Manufacturers like New United Motor Manufacturing, Inc. (NUMMI) issued uniforms, incentive pins, and calling cards to every employee in order to build loyalty and reduce barriers between managers and workers. The NUMMI motto: "The expert on the job is the worker on the floor" demonstrated management's new respect of hourly workers.

Self Serv Teller

delivers instant cash anytime you need it, 24 hours a day!

After banking hours, on weekends—even on holidays. No waiting in teller lines. The SELF-SERV TELLER is outside the bank under a bright canopy that is illuminated at all times. Your personal code number and special Master Charge card with its magnetic strip are all you need for this 24-hour, 365-day banking service. Nearby is our night deposit slot for depositing checks, cash, or loan payments at any time.

NOTE: Only you and Wells Fargo know your special code number, so no one else can use your card for SELF-SERV TELLER withdrawals.

here's how it works:

1 Insert your special Master Charge card in the slot.

2 Register your code number and amount of cash desired ($25 or $50) on the keyboard.

3 Collect your cash, Master Charge card and withdrawal slip. The withdrawal is automatically deducted from your checking account.

Self–Serv Teller, 1970. Wells Fargo introduced its first automated teller machines (ATMs) in 1970 in their San Diego central office. The machines allowed people to access their accounts and withdraw money any time, 24 hours a day.

NETWORKS

If computers and other smart devices are the building blocks of modern society, then communication networks are the mortar that holds them together. Sometimes networks were privately controlled (on leased telephone lines or radio waves), others were public, but all the networks gave businesses and consumers more choices, helped lower prices, and sped the pace of commerce. With the aid of personal computers (PCs), smart phones, and networked devices, people gained access to more information and products and, sadly, also became vulnerable to a new threat—cybercriminals. Some consumers had great optimism about the power of networks and hoped the Internet would fundamentally change the capitalist system and create a collaborative, moneyless exchange.

While Americans first experienced networks through proprietary systems like Automated Teller Machines (ATMs) in the 1970s and some businesses, like Walmart, installed satellite telecommunication systems in the mid-1980s, the widespread popularization of information technology took off with the Internet in the 1990s. After the creation of the World Wide Web in 1994, the use of PCs and dial-up modems gave everyday citizens access to information, shopping opportunities, and enabled social media.

One telling example of mobile devices and entrepreneurship coming together successfully involved a fusion food truck in Los Angeles, California, in 2008. Serving Korean BBQ with a chili-soy slaw in a tortilla—a Korean taco—Kogi BBQ was not immediately popular. However, when Kogi chef Roy Choi and the founders of Kogi began

FBI poster for Sun Kailiang, 2014. Cybercrime takes many different forms. Sometimes it involves hackers engaged in youthful vandalism; sometimes it is organized criminals seeking to defraud and gain access to other people's money. At its worst, cybercrime is state sponsored theft of industrial secrets and intellectual property.

iPhone, 2009. Kogi BBQ food truck chef Roy Choi used his phone to tweet information, find new locations, and chat with customers. In 2007 when Apple released its first iPhone, Steve Jobs described it as "a revolutionary and magical product." In this case the hyperbole may have been correct.

SAM WALTON
1918–1992

A child of the Great Depression, Sam Walton knew the value of a buck. Beginning in 1962, Sam Walton developed a discount retail chain, Walmart, aimed at middle- and working-class consumers living in rural communities. In 1983, he expanded the concept by adding Sam's Club, a membership-only retail warehouse chain featuring merchandise in bulk. With Walmart and Sam's Club, Walton brought low prices to customers through self-service, the removal of distributors, better management of his supply chain, and more. He was also an avid adopter of technology, using computers, telecommunication, and scanning to keep track of his inventory and customers' habits. Consumers at Walmart, attracted by its low prices, bought more goods, but critics worried that cheap goods lowered wages at home and abroad.

Hat worn by Sam Walton, about 1985.

building an Internet following by tweeting truck locations and chatting online, customers turned out. The company expanded and in 2014 operated four trucks and two sit-down restaurants. The old adage "if you build a better mousetrap, the world will beat a path to your door" is rarely true. Success in business takes a combination of good product, marketing, employees, location, and other variables. Many great businesses fail even though their concept is strong. In the case of Kogi BBQ, the use of tweeting (to gather and disseminate information) was a key to their business success.

The changes in the financial industry are even more remarkable that the success of Korean tacos. In the 1980s, the finance sector grew in size and importance as more average Americans entered the market through pension, 401K, and insurance funds. The speed and complexity of investing increased and regulation decreased. Professional money managers turned to riskier and complicated investments including derivatives and mortgage-backed securities, which they often did not completely understand. Investors were delighted by their portfolio growth but did not fully appreciate the extent of their risks.

Edmund J. O'Connor (1925–2011) was a financial leader and pioneer who helped computerize and change trading systems, valuing innovation over tradition. He started out as a commodity speculator at the Chicago Board of Trade—buying and selling mostly agricultural futures. Successful, he led the Board of Trade in the late 1970s away from traditional commodities, like soybeans and corn, and into bigger and more volatile derivative markets (such as the S&P 500 index and later mortgage-backed securities). O'Connor at the Chicago Board of Trade and Leo Melamed at the Chicago Mercantile Exchange then moved the commodities markets away from the open outcry system where traders had stood in octagonal pits buying and selling futures with shouts and hand signals to electronic trading that relied on computers networked together. The digitization of the markets in the late 1990s through computers on lightning-fast networks

Commodity trader jacket, worn by Edmund J. O'Connor, about 2000. Being seen by other traders was important in the open outcry commodity trading pits of the Chicago Board of Trade and physical size helped. Wearing his 44 long jacket, O'Connor stood out.

WARREN BUFFETT
1930–Present

Warren Buffett is known for simple roots and frugality. As a boy growing up in Omaha, Nebraska, he sold Wrigley gum and Coca-Cola door-to-door. In 2014, when he was worth billions of dollars, he drove an ordinary car and lived in the same house of over 50 years. But Buffet was far from average—he was personally driven and had unusual opportunities. At age 10, the New York Stock Exchange was a must-see stop on a visit to New York. At age 11 he had read every book about investing in the Omaha Public Library and he made his first stock trade. "I was wasting my life up until then." At age 12, the son of a Congressman, he was transplanted from Nebraska to Washington, D.C.

Warren Buffett, 1990.

opened the door for the use of mathematical-model decision-making and high-frequency trading. Life was never the same.

The twenty-first-century shift first to manual electronic trading and then to fully automatic trading was not limited to Chicago futures exchanges. Financial centers in New York, Paris, London, and around the world also changed. Data centers housing both trading firms' computer servers and the exchanges' matching engines became key. The human element remains critical however. On the afternoon of May 6, 2010, a "flash crash" (automated response selling programs) swept the automatic high-frequency trading and the Dow Jones Industrial Average (the DOW is an index of 30 publically traded stocks and considered the most important indicator of stock market change) temporarily dropped about 9 percent in value (about $1 billion at the time). A built-in emergency stop algorithm kicked in at 2:45 PM and halted all trading for 5 seconds—long enough for humans to assess the situation and realize that no world event was causing the markets to crash.

AGRICULTURE

Although food is not as big an economic sector as finance, farming became a big issue in the Global Era. Is the production of food harming the environment and consumers? Is the food supply safe? Can the food needs of a growing world population be met? These and many more questions made food and farming important everyday topics of discussion and concern.

The cost of food dropped as farmers and ranchers increased the scale of production and adopted new innovations from biotechnology and the Global Positioning System (GPS) to large-scale organic and sophisticated irrigation techniques. While some consumers were willing to pay higher prices for food that was grown on small and local farms because they thought it safer and tasted better, most Americans remained comfortable with the low cost and diverse selection of mass production.

Farmers and ranchers became increasingly dependent on new equipment, chemicals, and hybrids and in doing so took on increased financial risk. Education and environmental regulations improved agricultural practice, but reliance on the efficiencies of monoculture and commercial techniques continued to create worries.

One of the most contentious areas of discussion was around biotechnology. The story of Roundup Ready soybeans revealed the promises and challenges of the new technology. The new genetically modified seed, resistant to the herbicide Roundup, helped fuel a revolution in farming practice.

In the 1970s, the field of biotechnology took off as San Francisco scientists like Herbert Boyer and Stanley Cohen successfully transferred a gene from one species into another. A biotechnology craze then built as venture capitalist Robert Swanson teamed up with Boyer to form a company commercializing recombinant DNA (the manipulation of genetic makeup of organisms). Scientists at their company, Genentech Inc., engineered a bacterium that would produce the much-needed drug human insulin. The new manmade drug was an economic and social success. Herbert Boyer was even put on the cover of *Time* magazine's March 9, 1981, issue. At the same time, the United States Supreme Court ruled in *Diamond v. Chakrabarty* that a living organism created by humans could be patented.

Meanwhile, back in St. Louis, Ernest G. Jaworski, deputy to the chief scientist at Monsanto (a leading chemical company), hired Dr. Robert T. Fraley (1953–) in 1981. His job was to figure out how to get new genes into plant cells. The technique to actually insert new genes into a plant was very challenging. In 1983, Fraley along with Monsanto scientists Rob Horsch and Steve Rogers attended the annual Miami Winter Symposium on Molecular Genetics of Plants and Animals in Miami Beach and revealed to their peers how they had inserted the kanamycin-resistant gene into a petunia plant cell using the microbe *Agrobacterium tumefaciens*

Robert Fraley and the first transgenic petunias, about 1983. Born on a farm and educated in genetic engineering, Robert Fraley saw a new future for biotechnology–agriculture. As a pragmatic manager, he assembled and managed the massive teams of scientists it took to introduce herbicide resistance into petunias, and then soybeans.

as the delivery agent. While the new petunia had no functional value the experiment proved that transgenic or genetically modified plants with new immunity traits (kanamycin is an antibiotic that inhibits plant growth) were possible. Other scientists followed with different techniques for inserting new genes into plants.

The first gene gun (a different technique for introducing genes into plants) was the brainchild of John Sanford, a plant geneticist and assistant professor of Horticultural Sciences, Cornell University, in the early 1980s. The invention was a crude but ingenious concept: using the principles of a normal gun to blast DNA-covered microbullets at plant cells, thereby introducing foreign DNA

MARKETING IN THE DIGITAL AGE

Digital advertising began with wide public use of the Internet. An early, popular email provider called Hotwired began accepting advertising in 1994, and many other sites soon followed, often using pop-up or banner ads. Digital media transformed advertising's ability to speak directly to consumers and gather information on their behavior while replicating earlier advertising techniques of how to appeal to consumer desires. Advertisers continued to develop techniques that tied film, radio, print, television, and retail together, but digital advertising allowed them to gather more real-time information about consumers than previously possible. Much like the direct marketers before them, a new breed of digital advertisers followed consumers' digital footprints and tailored advertising to individual interests, needs, and desires. Taking an additional cue from narrowcasting on cable television, advertisers continued to define audiences and markets in narrower and narrower terms. Unlike television, however, the Internet could provide an almost instantaneous loop back to data miners about consumer habits, spurring debates about privacy and net neutrality. While some consumers saw the new digital advertising as invasive, others appreciated the convenience of seeing product and service information directly on their devices, allowing them to compare prices and scan consumer reviews. As of 2014, advertisers spent less on digital advertising than more traditional methods, but experiments and controversies continued as online advertising grew.

GEICO advertisement featuring Maxwell the Pig holding a smart phone, about 2010.

and creating genetically modified plants. As word of Sanford's research spread, genetic engineers at Agracetus, the Middleton, Wisconsin-based biotech firm, found their own inspiration.

Spurred by word of Sanford's gun, Agracetus employees Dennis McCabe and Brian Martinell invented a gene gun of their own in 1986. Cobbled together from scrapped radar station parts McCabe had purchased years earlier from the University of Iowa and potato chip bags from the Agracetus vending machine, the prototype, seen here, utilized a high-voltage electric shock to transform a water droplet into a shock wave that drove DNA-coated microparticles of gold into plant tissue. By April 1988, McCabe and Martinell had used the gun to create the first genetically transformed soybeans. Their success led to a deal with Monsanto to develop Roundup Ready soybeans.

Earlier, attitudes and direction at Monsanto had changed as the company, hit hard by the oil crisis, was deeply in debt. In 1986, the new CEO Richard Mahoney dramatically redirected Monsanto, selling off the petrochemical divisions and moving into biotechnology. Fundamental research (the type that allowed developing genetically modified petunias) gave way to a pragmatic product focus. Mahoney explained, "We are not in the business of the pursuit of knowledge; we are in the business of the pursuit of products" (Charles 2001).

In 1989, Asgrow (a seed company), Monsanto, and Agracetus decided to work together. Asgrow provided great soybean stock, Agracetus provided a gene gun, and Monsanto provided Roundup-resistant bacteria. Development work began on an herbicide-resistant soybean. But in the meantime other biotech products came out. Public reaction to genetically modified organisms (GMOs) was a worry for the biotech companies prompting them to go on public relations campaigns to quell fears about genetically modified foods. Not everyone was convinced however. In a 1992 letter to the *New York Times,* English professor Paul Lewis coined the pejorative description—Frankenfood. He wrote about GMOs "… ever since Mary Shelley's baron rolled his improved human out of the lab, scientists have been bringing just such good things to life. If they want to sell us Frankenfood, perhaps it's time to gather the villagers, light some torches and head to the castle" (MedicineNet). Persistent attitudes, like Lewis's, that conflate Shelley's fictional account built on primeval fears with scientifically documented and tested plants have been a problem for biotechnology companies.

Often more receptive to innovation, Americans seemed more at ease with the new technology of GMOs than Europeans. The Flavr Savr tomato, the first commercially grown genetically modified food to be brought to market, was approved by the Food and Drug Administration (FDA) in 1994. The tomato was designed by the biotechnology company Calgene to stay firmer and have a longer shelf life than regular tomatoes. Despite its flavor saver name, however, it was not very tasty and even more critically the plant was not very productive. The tomato was soon pulled from commercial production. In 1995 Asgrow also had a GMO squash; Monsanto had a GMO potato. It was not until 1996, though, that the real winner, Roundup Ready soybeans, was marketed. The new seeds allowed farmers to spray their plants in the fields, killing all the weeds and leaving the soybeans untouched.

Roundup Ready promotional souvenir, 1996. Herbicide-resistant soybeans were the first widely successful biotech crop. While expensive, farmers found the new seeds made their farming operations much more efficient as they simplified weed control. Monsanto's 1996 Annual Report, proclaimed, "In 1901, Monsanto was a start-up company. In 1996, we were again."

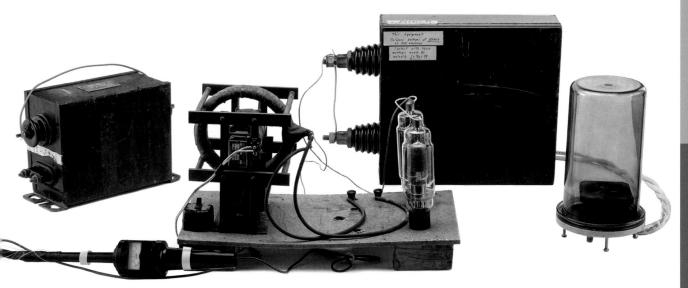

Gene gun, 1986. Solving a major hurdle in biotechnology, this prototype gene gun, developed by Agracetus employees Dennis McCabe and Brian Martinell, shot new genetic material into plant cells. Biotechnology quickly became one of the important topics for agricultural researchers.

Farmers hate weeds. They take nutrients and moisture out of the soil and block sunlight. In an agricultural community weeds also speak to character—a farmer with a weedy field is often looked down upon. In the past, farmers could plow or spray between rows but hand hoeing between soybean plants was left to children (walking the beans) or migrant workers (often undocumented immigrants). The introduction of Roundup Ready soybeans offered a solution to a major labor problem for farmers. In 2000, only four years after introduction, over 50 percent of American soybean acreage was GMO beans and by 2007 it was up to 91 percent. Later biotechnology was used to

introduce disease and drought resistance, and many other desired traits into a variety of plants.

The popularity of genetically engineered, herbicide-resistant plants led to an increase in the use of chemical sprays and that worried many people. But in an unexpected twist this trend in chemical use had positive environmental benefits. By using the new herbicide-resistant plants, farmers did not need to plow their fields to control weeds. Instead they could switch to conservation tillage practices like no-till where plowing is replaced by spraying. By not disturbing the soil, erosion dropped significantly in the United States. No-till also leaves more organic material like corn or

NO–TILL SAVES SOIL sign, mid-1980s. Seeking to promote environmentally friendly no-till conservation tillage through example, the Princeton, Illinois, Soil and Water Conservation District posted signs like this one on local farmer Jim Rapp's fields. With no-till, farmers use herbicides for weed control, not plows.

OUTBACK S GPS guidance, 2004. The Global Positioning System (GPS) revolutionized farming. Precision-guided equipment gathered crop yield information and decreased the amount of seed, fertilizer, and chemicals; farmers stopped thinking of their land as a single homogenous field to treat equally in terms of fertilizer, drainage, and herbicides.

wheat stubble in and on the ground, improves water absorption, and decreases evaporation. Even worms do better without the disruption of plowing.

Another huge component in the development of high-efficiency agriculture is the adaption of the Global Positioning System (GPS) to farming. Created by the U.S. Department of Defense in 1973, GPS is a U.S.–owned satellite-based utility that provides users with positioning, navigation, and timing services. The civilian signal was purposefully fuzzy until May 2000, when the government allowed civilian applications to be more accurate. In agriculture, the innovation was first used to guide tractors by keeping them on a straight line. Soon its use extended past just

vehicle guidance and was used for activating seeders, sprayers, and other devices. By applying seed, fertilizer, and pesticides more precisely, GPS-enabled devices reduced the environmental impact of raising crops and increased efficiency. GPS technology also allowed farmers to create accurate yield maps that show how their fields perform at specific points.

The story of organic food is equally fascinating. In the 1960s and 1970s, back-to-the land and organic farming was a cult movement, largely driven by counterculture concerns over pollution and the environment. Some urbanites turned to communes and small-scale agriculture as the idyllic life. Following the direction of visionary publications

MYRA AND DREW GOODMAN
Drew, 1961–Present
Myra, 1964–Present

New Yorkers Myra Rubin and Drew Goodman moved to California to go to college. There they fell in love with each other and organic salad greens. In 1984 after finishing school (Myra at Berkeley and Drew at Santa Cruz) they started an alternative backyard operation, selling pre-washed and bagged salad greens on 2.5 acres in Carmel Valley, California. Knowing the potential of the East Coast market, they grew their West Coast passion for organic greens, and the practicality of pre-washed packaging into a commercial success. Convincing other growers that organic production could be profitable and with the aid of Myra's inventive father, they grew the operation, Earthbound Farm, to 33,000 acres by 2009— the largest producer of organic greens.

Myra and Drew Goodman, about 1986.

like those from Rodale Press, people with little farming experience learned how to grow food organically and eat only "natural" foods. Of course, the definition of natural is quite contentious as most food eaten in the United States is not indigenous and those plants that are native to North America have been hybridized over years of selection or mechanical crosspollination. Wheat is native to Turkey, corn to Central America, and tomatoes to South America. All have been hybridized to such a degree that they hardly resemble their ancestor origin.

In the 1980s, life began to change as the cult fringe was replaced by entrepreneurs who scaled up commercial organic production. Responding to the public concerns of people who were scared to go shopping because of GMOs, pesticides, antibiotics, and other perceived problems, organic production took off in the 1990s and 2000s. Companies like Cascadian Farms (1972) and Earthbound Farm (1984) led the way. Soon retailers would see an opportunity as well. In 2006 Walmart, responding to Middle American desires, greatly expanded its organic offerings. Mainstream growers like Dole and Del Monte also saw the opportunity and commercial organic production expanded.

Taylor Farms sign, 2012. Bruce Taylor came from a long line of California lettuce producers. Responding to the growing interest in organic food, he scaled up his commercial production of organic salad greens. Because many of his employees are Spanish speaking, this sign is written in both Spanish and English.

TV ad still, 2008. Faced with increasing price competition from national poultry producers, Foster Farms, the largest West Coast producer, commissioned a humorous ad campaign that promoted their chicken by redefining the word natural. The campaign coined the term "plumping" to describe the widespread industry practice of saline water injection.

IMMIGRANT LABOR

The U.S. economy has long been powered by migrants and immigrants, but in the Global Era this reliance on a foreign-born population increased dramatically. In 1980, 6.2 percent of the U.S. population was foreign born and by 2010 that number ballooned to 13 percent (the all-time United States high was 14.8 percent in 1890 and the low 4.7 percent in 1970). The liberalization of immigration laws, most notably the 1965 Immigration and Nationality Act, also known as the Hart-Cellar, helped bring about the change. At the same time the countries of immigrant origin shifted away from Europe to Asia and Latin America. Despite deeply held public beliefs, less than half of the immigrants in the United States were on a path to citizenship as the United States allows relatively few people to apply for permanent residency. Over half of legal foreign-borns in the United States are students or temporary workers on nonimmigrant visas. Estimates of undocumented workers vary but probably hovered around 3 to 4 percent of the population. Combined, the immigrant populations provided a rich diversity in the United States, increasing innovation, depressing wages, lowering labor strife (immigrants tend to be compliant workers), and providing a pool of new entrepreneurs. The United States, once characterized as a melting pot of immigrants, is now described by scholars

BALBIR SINGH SODHI
1949–2001

The American Dream has long made the United States a magnet to immigrants from around the world. Some immigrants achieve success, others find discrimination, and some experience both. Escaping the religious persecution of Sikhs in his native India, Balbir Singh Sodhi immigrated to the United States in 1987, hoping to use his training in mechanical engineering to make a living. Instead, he drove a taxi for thirteen years in Los Angeles and then San Francisco. By 2000 he had saved enough money to open his own gas station in Mesa, Arizona. He did well helping fellow immigrants as they tried to get a start. Four days after the terrorist attack of September 11, 2001, Sodhi was shot and killed outside his station, mistaken for an Arab Muslim.

Balbir Singh's San Francisco taxi driver's license, 1996.

with the metaphor of the salad bowl. The goal of total assimilation is out and recognition of the value of difference is celebrated. No one wants their onion to taste like a tomato.

There are many specialty nonimmigrant visas that allow foreign citizens to live and often legally work in the United States. One driving force for coming to the United States is education. American higher education is greatly valued around the world and a large number of people come from abroad for university degrees. The F visa is used for those attending school in the United States, and it generally does not allow off-campus work. A much smaller number of people come on L visas, which are for intra-company transfers of executives or those holding specialized knowledge. H visas are very important and they sort into three groups H-1B, H-2A, and H-2B; the difference is skill and sector. An H-1B visa is for highly skilled individuals, like computer engineers, and the government only hands out a limited number of these. Competition for the visas by companies seeking to hire foreign workers is intense. In 2013 all 65,000 H-1B visas were given out in five days. An H-2A visa is for temporary agricultural workers, and the H-2B is for temporary unskilled workers. Mexico is the largest sending country for unskilled workers, with India the largest provider of skilled workers. More than a dozen other visas exist for specialized situations.

The story of Yogeeswaran Ganesan serves as an excellent example of higher education driving American innovation. The son of a senior manager at Tata Motors, Yogee became excited by biophysics while attending university in his hometown of Madras (now known as Chennai), India. After receiving his bachelor's degree, he was drawn to Rice University in Houston, Texas, because of the school's reputation for cutting-edge work in the field of nanotechnology. While it was a difficult decision to leave his family and the comfort of his culture, he was drawn to the American university system and the chance to study with the best and brightest in science.

Meanwhile, high-tech companies like computer chip manufacturer Intel were always looking for talent to give them a competitive edge. In Silicon Valley over 60 percent of the scientists and engineers are foreign born. After receiving his PhD, Yogee Ganesan was identified by Intel for his specialized knowledge in the critical field of semiconductor research, earning him a job and a visa.

above: **Yogeeswaran Ganesan and his mother, Bhavani Ganesan née Rajan, in Palani, Tamil Nadu, India, 1985.** The son of a senior manager in the finance department at Tata Motors in Jamshedpur, India, Yogee grew up wanting to be a scientist.

right: **Intel badge, about 2012.** With a PhD in materials science and with expertise in nanomechanics, interfacial fracture mechanics, and microfabrication, Yogeeswaran Ganesan has the specialized skills much sought after by semiconductor manufacturers. He was hired by Intel Corporation in 2011 after graduating from Rice University.

Department of Homeland Security
U.S. Citizenship and Immigration Services

I-797A, Notice of Action

THE UNITED STATES OF AMERICA

| RECEIPT NUMBER ▓▓▓▓ | CASE TYPE I129 |
| | PETITION FOR A NONIMMIGRANT WORKER |

| RECEIPT DATE April 11, 2011 | PRIORITY DATE | PETITIONER INTEL CORP |

| NOTICE DATE May 6, 2011 | PAGE 1 of 1 | BENEFICIARY A138 003 542 GANESAN, YOGEESWARAN |

D SCADOVA
FRAGOMEN DEL REY BERNSEN & LOEWY L
RE: INTEL CORP
3003 N CENTRAL AVE STE 700
PHOENIX AZ 85012

Notice Type: Approval Notice
Class: H1B
Valid from 10/01/2011 to 08/27/2014

The above petition and change of status have been approved. The status of the named foreign worker(s) in this classification is valid as indicated above. The foreign worker(s) can work for the petitioner, but only as detailed in the petition and for the period authorized. Any change in employment requires a new petition. Since this employment authorization stems from the filing of this petition, separate employment authorization documentation is not required. Please contact the IRS with any questions about tax withholding.

The petitioner should keep the upper portion of this notice. The lower portion should be given to the worker. He or she should keep the right part with his or her Form I-94, *Arrival-Departure Record*. This should be turned in with the I-94 when departing the U.S. The left part is for his or her records. A person granted a change of status who leaves the U.S. must normally obtain a visa in the new classification before returning. The left part can be used in applying for the new visa. If a visa is not required, he or she should present it, along with any other required documentation, when applying for reentry in this new classification at a port of entry or pre-flight inspection station. The petitioner may also file Form I-824, *Application for Action on an Approved Application or Petition*, with this office to request that we notify a consulate, port of entry, or pre-flight inspection office of this approval.

The approval of this visa petition does not in itself grant any immigration status and does not guarantee that the alien beneficiary will subsequently be found to be eligible for a visa, for admission to the United States, or for an extension, change, or adjustment of status.

THIS FORM IS NOT A VISA NOR MAY IT BE USED IN PLACE OF A VISA.

Please see the additional information on the back. You will be notified separately about any other cases you filed.
U.S. CITIZENSHIP & IMMIGRATION SVC
CALIFORNIA SERVICE CENTER
P. O. BOX 30111
LAGUNA NIGUEL CA 92607-0111
Customer Service Telephone: (800) 375-5283
Form I797A (Rev. 09/07/93)N

PLEASE TEAR OFF FORM I-94 PRINTED BELOW, AND STAPLE TO ORIGINAL I-94 IF AVAILABLE

Detach This Half for Personal Records

Receipt # ▓▓▓▓▓▓
I-94# 320814158 20
NAME GANESAN, YOGEESWARAN
CLASS H1B

VALID FROM 10/01/2011 UNTIL 08/27/2014

PETITIONER: INTEL CORP
2200 MISSION COLLEGE BLVD
SANTA CLARA CA 95054

320814158 20

Receipt Number ▓▓▓▓▓
Immigration and
Naturalization Service
I-94
Departure Record   Petitioner: INTEL CORP

14. Family Name GANESAN	
15. First (Given) Name YOGEESWARAN	16. Date of Birth 10/13/1981
17. Country of Citizenship INDIA	

Form I-797A (Rev. 10/31/05) N

I-797A Notice of Action, Department of Homeland Security, 2011. As a nonimmigrant worker, Yogeeswaran Ganesan was not on a path to citizenship. His ability to stay and work in the United States was contingent upon Intel sponsorship of his H-1B status. The I-797A form is the official notification of status.

The experiences of Dora Escobar are much more in line with the classic story of American immigration. Born in El Salvador, Dora got her first job at age six. She knew that work was the only way to achieve what she wanted. In1992, at age twenty-three, Dora journeyed to Los Angeles, seeking opportunity but finding exploitation.

Undaunted she pressed on. Moving to Maryland, Dora reunited with her husband who had emigrated five years earlier. Entrepreneurial and hardworking, she labored in the "underground economy," selling clothes and making a Salvadorian favorite—*pupusas*, thick tortillas filled with meat or cheese—in her home. Unlike earlier entrepreneurs such as John Rockefeller or Hattie Carnegie, Dora Escobar did not set off to build a large business and become wealthy. Instead she intended to just earn a living and "treat my customers well, to take

Dora Escobar, 2014. Standing in the kitchen of her Tacoma Park, Maryland, restaurant La Chiquita, Dora Escobar proudly holds a *pupusa* platter. Serving her community of immigrants, Escobar built several successful businesses. Her restaurants feature Guatemalan and Salvadorian foods and her check-cashing services provide a much-needed financial infrastructure.

U.S. HISPANIC MARKETING

National advertisers began to appreciate the value of the diverse and growing communities of Mexican American, Cuban, Puerto Rican, and other Latino consumers in the 1970s. Producers' and retailers' desire to reach the Hispanic customers provided an opportunity for Latin Americans to get jobs in the advertising industry and eventually open agencies focused on recognizing, defining, and selling to an often-undifferentiated Hispanic market beginning in the 1980s. The campaigns were carried through traditional U.S. print, radio, and TV channels as well as a growing number of Spanish-speaking media outlets. In the 1990s, Latino marketing gained power as a niche within the larger industry, as the Latino population in the United States increased in size and gained purchasing power.

Goya Foods sought both Spanish and English-speaking customers, 1960s–1990s.

RYUJI ISHII
1952–Present

Born and raised in Japan, Ryuji Ishii hated the restrictiveness of Japanese society. He came to the United States in the late 1970s, studied accounting, and then took a job coordinating the franchise department of an Asian American fast-food chain. Frustrated by the company's unwillingness to institute his reforms, he decided to go into business for himself in 1986. He combined his fascination with supermarket delicatessens, franchise knowledge, and memories of childhood food (sushi) to found Advanced Fresh Concepts (AFC) and the notion of supermarket sushi. Under his innovative business model, supermarkets provided the space, franchisees (Asian immigrants putting up little money) the labor, and AFC the training and infrastructure. All shared in the proceeds. Soon everyday Americans from Alabama to Alaska were eating sushi.

Ryuji Ishii at his Los Angeles headquarters sushi bar, 2014.

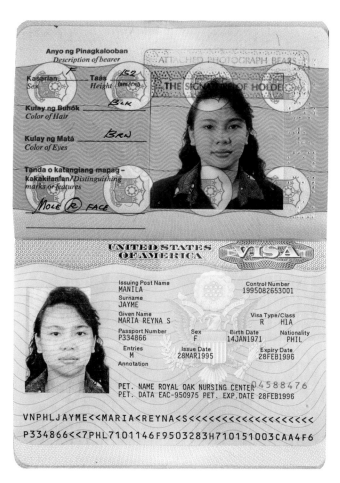

United States visa, 1995. Workers with highly desirable skills received special visas. To relieve a nursing shortage, Congress passed the Immigration Nursing Relief Act of 1989, giving special visas and eventually allowing a path to citizenship. Filipina nurses like Maria Jayme were highly valued because they spoke English.

care of what life gave me every day" (Diaz). Saving her money, Dora bought a used food truck and built a business selling *pupusas* to the émigré community. Fearful of debt, Dora slowly built her business with cash, expanding her offerings with three restaurants and 175 employees. Understanding the needs of the undocumented people in her community who often chose to stay off the grid and not open bank accounts, Dora opened nine check-cashing facilities. While many immigrants struggle to survive, others like Dora Escobar flourish as natural entrepreneurs building a future for themselves and their employees while serving their community.

RESEARCH AND DEVELOPMENT (R&D)

For big business, the balance between funding fundamental research and product development research is challenging; both drive innovation and ultimately economic growth. Some companies use research to push for new efficiencies in operations, others search for more sustainable choices, and many seek competitive advantage through new products and materials.

Around 1900, big business had begun to invest in research laboratories, using science to secure an edge on their competition. While some U.S. companies, like AT&T, DuPont, and RCA, had famous labs, most fundamental breakthroughs came from Europe until World War II. The U.S. innovation advantage lay with execution and taking ideas to market—technological development, manufacturing, marketing, engineering, and the size of the U.S. domestic market. During the Cold War period (1945–1980s), the U.S. government spent increasing amounts on basic research, funneling most support through university programs.

Money plays a critical role in research. In the Global Era, research continued to be a critical business decision, but the associated costs skyrocketed and funding sources began to shift. Federal money for fundamental research through military support had been common during the 1950s and 1960s but diminished as a percentage of the total spent on research in the 1980s. Since the early 1980s, U.S. universities have taken an increasing interest in entrepreneurial activities, association with venture capitalists, and commercially relevant research that can increase economic growth and promote technological progress. Industry itself has become more active in supporting leading-edge university work and then taking further development to their own labs. Interestingly, the weakening of antitrust laws and strengthening of intellectual property (IP) protection has changed the field, making industry generally more willing to allocate funds to research.

The location of research and techniques employed ranged from in-house labs and university collaborations to crowd sourcing and buying small companies with new ideas. The day of the lone inventor or even the entrepreneurial inventor (such as Samuel Morse with the telegraph) has dramatically declined. But the easy access to significant amounts of venture capital has greatly encouraged small start-ups, especially in the information technology sector.

In other areas, like physics and biotechnology, the problems now being tackled have technical demands that often require large teams of people with highly specialized knowledge working together. Quite often the challenge was creating an

3M Golden Step Award, Cubitron, 2011. Seeking to motivate its workforce, 3M created the Golden Step Award, which recognizes product innovation resulting in significant profits to the company. To qualify, the products have to generate at least $10 million in annual global sales within the three years of product introduction.

environment where people with wildly different backgrounds could get together and then motivating them to create the next breakthrough. Of course, all industry is not the same and the willingness of publically traded corporations, which need to show quarterly profits to fund long-term research, is sometimes less than that of privately held companies.

The rise and influence of management consulting firms affected the allocation of research funds by business. Just as management consultants promoted time-motion studies in the early 1900s, today major consulting firms promote ideas for increasing innovation. As companies get bigger, individual managers' confidence in decision-making weakens and the need for an outside group to legitimize decisions has increased. Four major management consultant companies dominated American corporations: McKinsey & Company; the Boston Consulting Group; Bain & Company; and Strategy& (formerly Booz & Company). When asked, "Is America as innovative as it used to be?" these companies have a vested interest in being concerned. After all, if America is still very innovative, then a company doesn't need to hire a management firm to fix the problem because there is no problem.

Other major considerations in promoting U.S. inovation are higher education and diversity. With high-quality universities coupled to the promise of leading-edge work and social mobility, the United States remains an attractive destination for international students. Diversity in the workplace has been a major factor to American success. The Internet and computer technology have made the world smaller, but many very smart people still make the difficult decision to leave their families and homelands to study and then work in the United States.

Innovation does not follow a single course, so a prescriptive approach to inculcating innovation is nearly impossible. Putting a foosball table in the break room, for example, may just be a distraction— not a meeting point that creates change. Two case studies that provide considerable insight into the mangagement of innovation are sequencing the cacao genome and 3-D transistors. Sequencing the cacao genome followed the course of a collaborative dispersed global team and an open source result. The development of 3-D transistors followed a different track—the proprietary lab. There are, of course, many other successful models.

Harvesting cacao, Ghana, 2011. Cacao (from which chocolate is made) is mostly grown by small land holders using simple techniques. At harvest time the pods are cut and the seeds are scooped out. The pulpy material is allowed to ferment for a few days and then the seeds are dried.

Cacao trees, the source of all chocolate, are grown in equatorial regions of Africa, South America, and Asia by about 6.5 million small land owners using simple techniques. But the trees are endangered by diseases such as black pod, witches' broom, and frosty pod rot. Additionally, the low productivity of the trees helps lock farmers into poverty. Unfortunately, agribusinesses are not motivated to spend large amounts of money to do the research that would enable them to come up with new hybrid trees or chemicals to fight diseases because the farmers are too poor to pay for the improvements. Additionally few of the governments in production regions have put money into research.

In 2008, fearing a long-term collapse of cacao production, candy producer Mars, Inc. funded fundamental plant research to sequence the cacao genome. The research provided the first step to developing disease-resistant and more productive trees. Instead of assigning the work to a company lab and keeping the results to themselves protected by patent restrictions, the company chose a different strategy. Mars funded a $10 million collaboration, including the USDA, IBM, the National Center for Genome Research, Clemson University, Hudson Alpha Institute for Biotechnology, Indiana University, and Washington State University. After they succeeded in sequencing the genome, they made the research open source, giving the results away to anyone who was interested, hoping that others would build upon the information and hybridize new disease-resistant cultivars.

Genome sequencing is the process that uncovers the unique DNA code of a species. DNA, the molecule of genetic inheritance, codes for life using four nucleotide bases, represented by A, G, T, and C. In cacao, there are about 420 million DNA units. The order of these bases in an organism's DNA provides the instructions to carry out life's functions. A sequenced genome, therefore, provides an important steppingstone for understanding the genetic basis for variation in a species.

In the case of sequencing the cacao genome, researchers in Costa Rica harvested cacao leaves, pods, and flowers and sent them to a USDA lab in Miami. USDA scientists extracted genetic material from the cacao samples and sent it to university labs around the United States for sequencing. DNA samples from the cacao were primarily analyzed

Howard Shapiro, Ghana, 2009. Standing beside a cacao tree in Ghana, Howard Shapiro, Global Director, Plant Science and External Research, Mars, Inc., carried out field research. In 2007, he assembled and directed a worldwide team of scientists to sequence the cacao genome. Effective leaders need both management and technical expertise.

on a second-generation Roche 454 sequencer with gaps filled by older Sanger sequencing. The sequence apparatus only read a small portion of the DNA molecule, so a tremendous number of runs had to be pieced together and a massive amount of data interpreted. In 2010 (three years ahead of its scheduled end date of 2013) the collaborative team met with success and sequencing of the cacao genome was complete.

Knowing the genic structure of cacao greatly speeds up the hybridizing process. While the creation of new cultivars that might be disease resistant or more productive is still done through traditional hand pollination, biotechnology allows scientists to test new plants quickly, without having to wait years for them to mature. Additionally, Mars made the results of the sequencing public (open access) and not a proprietary secret. This

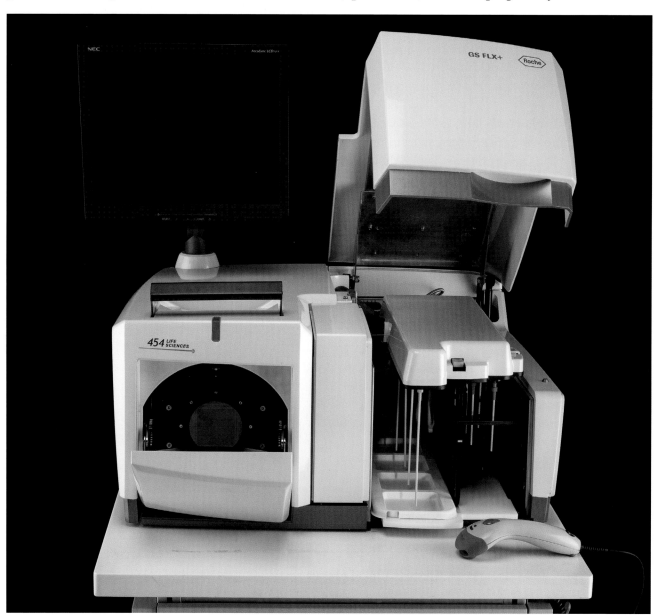

Roche 454 sequencer. The workhorse of the cacao project was the Roche 454. Dr. Keithanne Mockaitis, Director of Sequencing, Center for Genomic and Bioinfromatics, Indiana University, led a team of scientists in her lab sequencing the 350 million base pairs of *Theobroma cacao*.

2008 Achievement Award. The ten Intel achievement award winners were part of a team of hundreds of scientists. From left to right, front to back: First row: Kelin Kuhn, Robert Chau, Tahir Ghani. Second row: Brian Doyle, Ravi Pillarisetty, Matthew Metz. Third row: Michael Mayberry, Jack Kavalieros. Fourth row: Titash Rakshit, Uday Shah. Not shown is Annalisa Cappellani.

approach allowed hybridizers around the world with no connection to the company to take advantage of the information and build upon each other's work. Cross-sector scientific collaboration and open access is seen by many as a powerful approach to research, although the direct beneficiaries of the innovations enabled by such research are not necessarily the funders. Mars hopes to get a stable cacao supply eventually.

There are many successful models to research and develop. A different road to managing research success was taken by computer chip manufacturer Intel in developing a 3-D transistor. Driven by the need to make computer chips smaller and use less power, the company looked to university labs for leading-edge ideas.

In the 1990s, with the help of funding from the Defense Advanced Research Projects Agency (DARPA), university academics developed a revolutionary concept in semiconductor design— the move from planar to 3-D transistors. Intel thought that the theoretical idea might be commercially viable and began proof of concept research in their in-house Hillsboro, Oregon, research laboratory. Working in secret, it took Intel scientists six years to perfect the Tri-Gate chip.

Experimental transistor, 1948. Transistors, the building blocks of the digital revolution, were developed in the Bell Labs in 1947. Making transistors ever smaller and cheaper has been a key to the functional improvements and cost reductions of computers.

left, above and below: **Computer chip and micrograph detail, 2011.** This integrated circuit, about the size of a nickel, has over 1.4 billion transistors. In 2011, Intel was the first semiconductor company to use revolutionary 3-D transistors. Smaller and using less power, the innovation kept the company ahead of competitors for a few years.

Managing an innovation project is very challenging and requires different styles at different points. With Tri-Gate there were three distinct phases—research, development, and manufacturing—each overlapped with the others and each had different types of scientists and engineers involved. Since most early work fails, that work had to been done cheaply and received relatively little support from the company. In the beginning resources were slim and projects competed with each other. A sense of community built as a competitor on one project may be a collaborater on another. Engineers like Brian Doyle, who was involved in the very beginning of Tri-Gate, often had to beg for help from his peers. He explained: "a bitter rival for resources in one project one year can quickly become your close partner and bosom buddy on another project" (Doyle, pers. comm.).

At Intel the focus had first been on developing a Hi-K metal gate transistor and much money and staff time were directed to the project. The small number of people working on the 3-D transistor concept had to hustle for resources. Engineers like Doyle had to become entrepreneurial and creative to redirect resource to their project. Intel engineer Uday Shah recalls Brian's creative hustle: "He even resorted to illegal means of posting signs in the Fab [the clean room where computer chips are manufactured] with a photo of the HiK Metal Gate engineer with a line across his face—indicating *please don't help him, help me instead*! Tri-Gate in 2003 was like a start-up" (Shah, pers. comm.). Having great managers that can build this kind of camaraderie is in fact what often separated the successful companies from the rest of the pack. Sheer brilliance is rarely enough.

When Intel senior management became confident that the 3-D transistor could be profitably manufactured, more resources were allocated for development. Management paralyzed by uncertainty leads to failure but swashbuckling confidence is dangerous as well: making the wrong decision can spell financial ruin. While life became easier in some respects for the Tri-Gate engineers the days became much longer and the pressure more intense as the company "bet the farm" on the new idea.

Attracting and motivating the scientists and engineers who develop new technology are a challenge for any business. Good pay, the chance to work on the leading edge of technology, and a focus on team, not individual achievement, are all important. A great team is often made up of young bright people who don't know better. They need a passion to prove people wrong and support from

WALLACE H. COULTER
1913–1998

Some successful American business leaders are significant as both innovators and philanthropists. In the 1940s, Wallace Coulter invented a new way of counting and sizing small particles suspended in fluid. Patented in 1953, this technique was incorporated into the Coulter Counter that automated the previously labor-intensive process of calculating complete blood counts. A prolific inventor with 85 patents, he built a successful company revolutionizing the medical testing field.

Coulter lived a very simple life and in 1997 after selling the company, he applied his energy and fortune to the Wallace H. Coulter Foundation. The foundation carried on his lifetime passion for biomedical engineering and translational research— aimed at applying scientific discoveries to enhancing human health and well-being.

Wallace Coulter, about 1970.

managers to do their work or, as some say, "air cover" from senior management. Innovators tend to be resilient, self-driven, and excited about being first and smartest.

As a manager, Mike Mayberry, director of components research, knew well that the key to innovation was assembling a winning team. "No one really works in isolation... Our research and development engineers work with suppliers... We work with universities on very novel ideas and hire the best students" (Intel 2011). Brian Doyle elaborated: "We no longer live in an age where a lone researcher working single-handedly in a laboratory can do this. Although an idea can bubble up from a single mind and be captured in a patent, in order to make that idea a reality, it requires more than a squad of workers, more than a platoon, or even a battalion. It involves an army of literally thousands whose individual jobs in researching, developing, fabricating, designing, simulating and modeling, testing, laying out structures, and putting together the architecture of the chip make the final product possible" (Doyle, pers. comm.).

EARNINGS AND DEBT

Beginning in the early 1980s, income in the United States began to polarize between relatively few in number high-wage knowledge workers and a growing number of low-wage retail and service workers. Well-paid blue-collar jobs, which had once defined the American middle class, became increasingly scarce. Much of the change was driven by the new economic trends of the Global Era and earlier questions about the division of wealth returned. Was a new aristocracy emerging as the difference between rich and the bulk of society approached the gap last seen in the Roaring Twenties? Who should benefit from rising efficiency and economic growth? Should profits primarily go to the financiers and managers who provided capital and took risks? What was the proper balance between sweat and wit?

In the late 1970s, wages stagnated and inflation raged. Around 1985, as buying power dropped,

household debt in the United States began to increase faster than income. Not appreciating the risk they were accepting, many people borrowed to maintain a middle-class life—home ownership, retirement security, access to healthcare, college education for children, a car per adult, and family vacations. Increased debt helped fuel the U.S. economy but hurt some people. Easy access to credit cards, increased use of student debt, and complicated adjustable-rate mortgages with low introductory teaser rates worked well for some, and buried others in a pile of debt. When a cyclical downturn hit the economy in 2008, home mortgage debt, personal debt, and student loans pushed many people into crisis.

Home ownership has long been a goal of most Americans, and in the 1990s the U.S. government pushed banks to make home ownership accessible to less affluent Americans. At the same time, the financial industry successfully pushed for deregulation. Liberalization of banking rules in the 1980s and 1990s resulted in reorganization of the financial sector and new loan products. An important change was the Garn–St. Germain

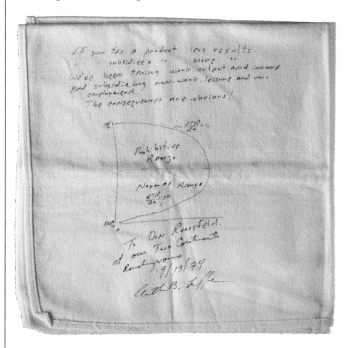

Napkin on which economist Art Laffer drew his sketch promoting the virtue of cutting taxes, 1974.

Depository Institutions Act of 1982, which allowed banks to offer adjustable-rate mortgages (ARMs). With small monthly payments, these loans made borrowing large amounts of money seem easy and helped fuel an unsustainable rise in real estate prices. The number of high-risk subprime mortgages increased from 5 percent of the U.S. loan market in 1994 to 20 percent in 2006. Lax oversight by loan underwriters sometimes enabled unqualified borrowers to take on large debt without background checks. Behind the scenes, the passage of the Financial Services Modernization Act of 1999 rolled back much of the New Deal era regulation imposed by the Glass-Steagall Act of 1933. The new law blurred the line between commercial banks and securities firms and enabled large amounts of money to be pumped into the home mortgage market, often funding high-risk loans. Like pouring fuel on a fire, extensive Wall Street investing in mortgage-backed securities promoted high-risk borrowing by unqualified home buyers. A perfect storm of bad choices by consumers of debt and producers of debt fed the housing market which skyrocketed to unsustainable levels.

For some, the availability of creative mortgages was an opportunity for the elusive American Dream to come true. A booming real estate market in the late 1990s turned into a bubble (from 2001 to 2006) with some people flipping homes for quick profits. Loan brokers became lax about documentation and checking for accuracy as they chased commissions. Most banks ceased to actually hold the loans as the stock market snatched up mortgage-backed securities even as rating agencies turned a blind eye to reality. Between 2006 and 2008, the bubble burst as low teaser rate ARMs reset and many loans became unaffordable for the borrowers. The value of homes dropped significantly. Foreclosures skyrocketed, driving down prices, and many home owners sank "underwater," owing more to their mortgage lenders than the market value of their homes.

JUDE WANNISKI
1936–2005

A popular author and the associate editor for the *Wall Street Journal*, Jude Wanniski helped form the modern direction of the Republican Party. In 1974, four men—Arthur Laffer, Dick Cheney, Donald Rumsfeld, and Wanniski—got together at the Two Continents restaurant in Washington, D.C. The topic was displeasure at President Ford's Whip Inflation Now (WIN) policy to lower inflation by raising taxes. To emphasize his point of view, economist Laffer drew on a restaurant napkin, arguing an old theory that lower taxes would increase economic activity and tax revenues.

Four years later, Wanniski recalled the incident, and gave it a catchy name. In an article entitled "Taxes, Revenues, and the 'Laffer Curve'," the sketch on a napkin took on new significance.

Jude Wanniski, about 1978.

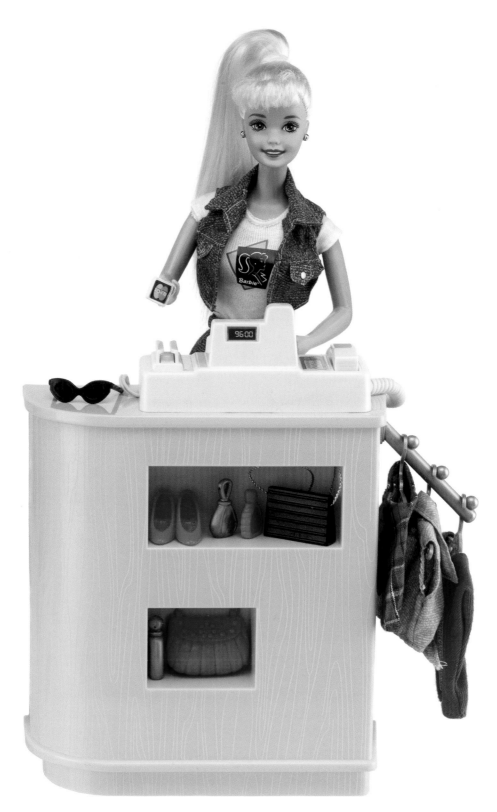

Cool Shoppin' Barbie, 1997. The toy industry helped prepare children for participation in the credit economy. Mattel Inc. and MasterCard worked together to develop a co-branded Barbie who operates a clothing store. When kids pressed the cash register button on this model, Barbie said "credit approved."

"Mortar Boarded," by Christopher Weyant, *The Hill*, May 30, 2012. A college degree is no longer a guarantee of a bright future. In 2011, nearly 20 percent of U.S. households owed a total of about one trillion dollars in student loans, far more than either consumer credit card or auto debt.

Personal debt as well as mortgage debt became a problem for many people in the Global Era. Short-term personal loans for consumer durables in the United States took off in the 1920s and increased in the Global Era. During times of rising wages or inflation, borrowing money is often astute or what economists call "productive debt." Borrowing money can be a form of savings if the object has economic value. During the Global Era, however, many consumers used credit to simply maintain their middle-class lifestyle, even as their income remained flat or declined. Banks promoted home equity lines of credit, loans that many consumers used to finance everyday purchases by extracting the equity on the inflated prices of their homes. The Tax Reform Act of 1986, which eliminated the deductibility of interest paid on credit cards, made borrowing against one's home even more desirable. Purchases ranged from jewelry and clothing to holiday presents and vacations. Debt can, of course, have bad effects. When someone is deeply in debt, even a small disruption (medical problems, unemployment, a car breaking down) can have a major ripple effect ending in bankruptcy.

Credit cards with revolving credit allowed consumers to pay only a small portion of the balance. From 1983 to 2008 credit card volume went up twentyfold, from $100 billion to $2 trillion.

Personal bankruptcy rates rose dramatically starting in the 1980s. The bankruptcy rates in the

Walt Disney World souvenir Mickey Mouse ear hat, 1996. For many people a family vacation was part of the middle-class American Dream. While some saved to go on their dream trip to destinations like Disney World in Orlando, Florida, others financed their trips with borrowed money.

United States stayed steady from 1900 to 1960, when they began to increase slightly. As of 2004, the filing rate was 5.3 per 1,000 people, more than four times the 1980 rate and 80 times the 1920 rate.

A third troubling area of indebtedness was education. Between 1985 and 2012 the cost of education rose three times faster than inflation. At the same time, loans to students became more widely available. Students took on unprecedented levels of debt, creating what some feared would be an "indentured generation." As the economy stalled and salaries stagnated, paying off student loans kept many graduates from beginning new households at the same ages as their parents and thus slowed their contributions to general economic growth. This form of debt also had a sobering proviso that often went unnoticed: student loans could not be discharged through bankruptcy.

Americans' economic struggles refocused attention on whether the nation remained committed to balancing individual opportunity with protecting the common good. Beginning in the 1980s, increasing income inequality shook American's belief in social mobility, a strong middle class, and for some the future of American capitalism. Throughout the Global Era total U.S. income increased, but more than 80 percent of the growth went to the top 1 percent. Popularized in the work of economists Thomas Piketty (author of *Capital in the Twenty-First Century*) and Emmanuel Saez, the public gaze shifted from focusing on the who and why of median income earners to examining the super-rich. In September 2011 emotions ran high on the streets in the financial district of New York City as angry Occupy Wall Street protesters demanded change. With no clear-cut demands, angst over corporate influence and the widening split of the profits of productivity took the form of public protest that soon swept the country.

Piketty worries that "the egalitarian pioneer ideal has faded into oblivion and the New World may be on the verge of becoming the Old Europe of the twenty-first century's globalized economy." Columnist John Cassidy adds that "since 2009, corporate profits, dividend payouts, and the stock market have all risen sharply, but wages have barely budged." However Piketty's perspective ignores the possibility that efficiency in production has made the cost of goods cheaper and raised the standard of living even for an expanded class of poorly paid workers. While the division of wealth might not be fair, the real cost of survival (food and shelter) in 1928 was lower in 2012 than 1928.

The past may provide some important insights as to what lies ahead. History shows that the United States has generally followed a course of continuous, but often disruptive change. During the period from 1900 to 1935 (like 1980 to 2015) the incomes between the rich and poor widened. Then, from 1945 to 1970, wages converged in what is often called the Golden Age. Many factors went into creating this period of "great compression"—strong unions fighting for wages and benefits, innovations making workers more productive, government regulating the power of corporations, and the overall strength of manufacturing during a time when industry around the world had been destroyed.

In the Global Era, the question of how American capitalism should distribute its wealth was hotly contested. Political economist and professor, former Secretary of Labor Robert Reich argued, "The liberal idea is that everyone should have fair access and fair opportunity. This is not equality of result. It's equality of opportunity" (Konformist 2004). Others, like the laissez-faire economist and professor George Reisman, unabashedly argued "...capital accumulation and economic progress depend on saving and innovation and that these in turn depend on the freedom to make high profits and accumulate great wealth" (1998).

Foreclosure sign, 2012. This foreclosure sign hung on a home in Fort Lauderdale, Florida. During the financial crisis of 2008, a triple whammy led to a mammoth rise in home foreclosures—people lost jobs, low teaser-rate adjustable-rate mortgages reset, and the housing bubble burst.

Can the World Work Together to Address Climate Change?

HENRY M. PAULSON JR.

Hank Paulson, former Treasury Secretary and CEO of Goldman Sachs, has devoted his life to solving complex financial problems. Now as chairman of the Paulson Institute, a nonprofit working to strengthen ties between the United States and China to address global challenges, one of his key priorities is to raise awareness of the looming climate change crisis. A lifelong conservationist, Paulson has been chairman of The Nature Conservancy and has studied the effects of global warming on the environment. He believes climate change poses significant economic risks to the United States and other major economies and requires immediate action.

Globalization presents broad benefits and opportunities, as well as enormous challenges, as businesses and governments wrestle with economic, social, and environmental change. Indeed, the benefits are realized through the advancements of societies and expanded opportunities for citizens around the world. But along with this progress come immense challenges that can affect each of us, no matter which country we live in. And what makes these issues particularly difficult to solve is that they often require nations to work together, which presents a challenge of its own.

Climate change is a case in point. I can think of no threat that poses a greater risk to our environment, our economy, or the quality of life for future generations. Given the mega-challenge that this issue presents, I am hopeful that it will force nations to work together to solve this problem. This will require heavy lifting by many nations, but especially the United States and China, the world's two biggest economies and energy producers. Our ability to work together on climate change will be a critical litmus test and could help shape the future of this most-important relationship in the decades ahead.

To meet the challenge of climate change, governments and businesses alike must assess both the short- and long-term risks and act before the economic and environmental consequences of failure are irreversible.

In an effort to evaluate the risks that climate change poses to the economy here in the United States, in 2013 I joined with former mayor of New York City Michael R. Bloomberg and philanthropist Tom Steyer to co-chair the Risky Business Project. We took a granular look across regions of the country and key economic sectors—and the results were truly sobering.

The U.S. economy faces multiple and varied risks from unmitigated climate change. These are disproportionately significant in certain regions of the country, and they are not all decades in the future: For example, projected changes in sea level, combined with changes in hurricane activity, will likely increase the cost of coastal storms along the East Coast and Gulf of Mexico by 28 percent to 60 percent over the next 5 to 25 years, or an additional $6 to $13 billion in average annual damage.

In the Midwest region, some states, including my home state of Illinois, will likely experience significant losses in crop yields for our major commodity crops of corn, soy, wheat, and cotton. Absent major adaptation efforts on the part of farmers and agribusiness, some states in the Southeast, lower Great Plains, and Midwest risk up to a 50 percent to 70 percent loss in average annual crop yields by the end of this century.

And for states across the South, hotter conditions will make outdoor work nearly impossible for large swaths of the summer. Texas, for instance, has experienced an average of 43 days per year with temperatures above 95°F over the past three decades. This number will likely nearly double between 2020 and 2039.

These trends require the private sector to manage climate risk, and we are beginning to see businesses adapting. Colgate has closed or relocated almost sixty manufacturing sites around the world as part of a restructuring program, focusing on sites increasingly exposed to severe weather conditions. Companies are also beginning to make infrastructure investment and siting decisions based on the science of climate. Shell Oil, for instance, employs an advisor to assess the future climate change conditions for large new projects in regions such as the Arctic, the North Sea, and tropical areas.

We can't stop at adaptation. It's important that companies incorporate climate risk into their business decision-making processes. Investors, as well, should and will increasingly demand that companies disclose their climate risk in financial reporting, which would sharpen the focus for management and investors. And the business sector is called to take a more active role in encouraging government to put in place the policy framework necessary to ensure a more sustainable economic future.

Climate change also poses a huge fiscal risk to our country. It's frightening, but not surprising that lawmakers often either relegate climate change to the back burner to address seemingly more immediate issues or dismiss the topic on political grounds. But government has a responsibility and an incentive to take the long view.

When natural disasters strike, government intervenes, spending billions of taxpayer dollars in disaster relief and recovery, and to shore up infrastructure. This is the proper role of government, but policymakers can't afford to continue to ignore the underlying issue.

The federal government should instead be assessing the fiscal realities of inaction. It should significantly increase basic research on new clean energy technologies, which only the public sector can do, and put policies in place that let the market direct resources toward investments that help reduce carbon emissions. A price on carbon, for instance, would help unleash a wave of innovation for new technologies, drive efficiencies, and change corporate and consumer behaviors.

Stemming this issue requires a global full-court press, which needs strong leadership from both the United States and China. The climate deal struck by President Obama and President Xi at the November 2014 APEC Summit in Beijing is an important and commendable step in this effort. Frankly, continuing to work closely with China may be our only real hope for solving the climate crisis. This is one of the areas where our countries' private sectors, governments, and nonprofit institutions have a strong shared interest in working together and in complementary ways to push for actions and to develop new clean technologies that can be deployed on a cost-effective basis in the developing world.

The challenge will be the speed with which we can come together. But the good news is that no nation on earth innovates better than the United States, and China can roll out and test new clean energy technologies on a speed and scale like no other.

For its part, China's air quality has reached a crisis point, and the government has no choice but to act. Spend a day in Beijing, which suffered some sixty days of pollution at hazardous levels in 2013, and it's no wonder Premier Li Keqiang has declared a war on pollution and launched a plan for economic reform to set China on a more sustainable environmental path. For instance, the Chinese government has introduced performance indicators for officials based not only on economic performance and social stability, but also on environmental management and quality of growth.

China is also taking steps toward pricing greenhouse gas emissions. Seven regional pilot carbon markets have been up and running in major cities since 2013, with the goal of developing a model for the country.

These are commendable efforts, but China has been losing ground from the impacts of breakneck growth that have overwhelmed the economy at a significant environmental cost. Chinese citizens—and the rest of the world—will demand they do more.

To meet the targets laid out at the November 2014 APEC Summit, the United States must do more to get our house in order by enacting policies to curb and price carbon emissions. When our own house is in order, we are in a better position to press China and other developing countries to take difficult but necessary steps to curb the crisis.

Given the stakes for future generations, the United States and the world community cannot afford to ignore the climate change crisis and its effects on our environment and our economy. It's as if we're watching as we fly in slow motion toward a mountain. We can see the crash coming, but we're sitting on our hands instead of altering course.

Does "American Enterprise" Exist in the Globalism Era?

RICHARD L. TRUMKA

Richard Trumka, president of the AFL-CIO, the nation's largest labor organization, has led a life that mirrors many of the changes that have swept through American business over the past half-century. Born and raised in Pennsylvania coal country, Trumka went to work in the mines in 1968 when he was nineteen. At that point, he worked for a U.S.-based company producing a product for domestic customers—circumstances common to most workers and companies back then. Today, Trumka presides over an organization that confronts global issues on a daily basis.

Any discussion of "American" enterprise in the current "Globalism Era" has to begin with a definition of terms. Can we really consider the large global U.S. corporation to be "American"?

If we want to chart a course for broadly shared economic prosperity in this era, we need a new understanding of the relationship between American democracy and the global corporation. There used to be no question that major U.S. corporations were "American." When the CEO of General Motors essentially said that what's good for GM is good for America, much of America agreed. From the end of World War II to the 1970s, America's businesses and workers prospered together. In fact, the incomes of the poorest 20 percent grew faster than the incomes of the richest 20 percent. But things were not perfect back then. America struggled with racism, sexism, homophobia, and other social ills. But the success of U.S. corporations with an unmistakable American identity coincided with unprecedented U.S. economic growth, broad-based wage growth, and the greatest reduction of economic inequality in our history.

Today, global corporations based in the United States are less likely to have an obviously "American" identity. Since the '70s, U.S. corporations have extended their global reach by moving assets and production overseas and expanding foreign markets and supplier networks. In 2014, many of the most profitable businesses headquartered in America, such as Google, have relatively small American workforces and largely avoid paying the U.S. corporate tax rate of 35 percent.

Even firms with deep American roots, such as General Electric and Ford, now organize their production globally. A light bulb or car may be designed in one country, while its components are sourced from others, it is assembled in yet another, and the finished product is marketed across the world. These corporations recruit their shareholders, lenders, and managers globally.

The global corporation has stopped investing not only in America, but also in its own future. Instead of investing in their employees, global companies increasingly have used their earnings to prop up stock prices to deliver cash to investors. From 2003 to 2013, companies on the Standard & Poor's 500 (S&P 500) used 54 percent of their earnings to buy their own stock (Lazonick, *Harvard Business Review*, September 2014) and another 37 percent of their earnings to pay dividends to investors, leaving a grand total of 9 percent for everything else. Many companies actually spent more money buying back stock than they earned, using borrowed money and essentially disinvesting in themselves.

Global corporations' own self-conception has changed, and a chasm has opened up between their perceived self-interest and the interests of the people who live and work in America. Former Colgate-Palmolive CFO Cyrill Stewart once said, "There is no mind-set here that puts this country first" (Balk 1990, 41). Cisco CEO John Chambers has gone further: "What we are trying to do is outline an entire strategy of becoming a Chinese company" (Prestowitz 2010, 212–13).

The growing divergence of interests between global corporations and the people who live and work in America was perfectly illustrated by the former CEO of General Electric, Jack Welch, when he said in 1998: "Ideally you'd have every plant you own on a barge to move with currencies and changes in the economy" (*The Economist*, January 19, 2013). A factory on a barge has no American identity and no sense of obligation or responsibility to the people of America.

The great contradiction of the global firm is that it scours the world looking for ways to profit from rock-bottom wages and tax rates, while at the same time expecting quality infrastructure, strong public education, and stable governments. The executives of these firms give speeches urging politicians to deliver all of these things, but it is not realistic to expect competitive infrastructure and education when they are keeping tax revenues low.

A serious problem arises when we ignore the plain reality that these corporations have little sense of responsibility to the people who live and work in America—yet we allow them to dominate U.S. politics, our democracy, and economic decision-making. Ironically, as our major corporations have chosen to be less tied to our national community, their power over our political and legal systems has grown, abetted by Supreme Court decisions like *Buckley v. Valeo* and *Citizens United*.

An instructive example is the policy area where these corporations have exercised the greatest amount of effectively unchallenged influence: trade. The Globalism Era is the only time in our history when we have run chronic and large structural trade deficits, which represent lost U.S. jobs. The end result of 40 years of trade policy designed by global corporations has been the systematic dismantling of U.S. manufacturing capacity, once the cornerstone of America's global dominance. Manufacturing, the business of making things, now constitutes a smaller portion of our economy than it does in the economies of our major competitors, such as Germany, South Korea, and Japan.

It is no coincidence that the Globalism Era has coincided with the end of shared prosperity, the stagnation of workers' wages, and the explosion of economic inequality. Whereas wages grew at the same pace as workers' productivity throughout the postwar period, the link between productivity and pay was broken (Tankersley, Wonkblog, July 17, 2013) toward the end of the 1970s.

In the years since then, workers' productivity continued to rise, but wages did not, and personal debt mushroomed. Our economy has been inflated by bubbles and rocked when those bubbles burst. Between 2010 and 2013, as CEO pay increased 21.7 percent, real wages *fell* 1.3 percent for private-sector workers. Inequality has reached staggering proportions—worse than any time in the last century.

These worrisome developments raise an urgent question: What role in setting our economic policies should we allow global companies to play when their interests are disconnected from those of the people who live and work in America? If America is going to prosper in the Globalism Era, we need to reach a new understanding with these global corporations: Either you are an "American Enterprise"—meaning you are committed to America—or you are not. Either you are on the team, or you are off the team.

Companies that choose to be on the "American" team must act as responsible corporate citizens: pay their taxes, follow rather than subvert our laws and regulations, invest in their own future, and support public investments. They can choose to produce in other countries, but they cannot pit U.S. workers against foreign workers in a race to the bottom, or defund America's infrastructure and educational system by conspiring with tax havens to minimize their tax liability. Companies on the "American" team deserve to be in the room when policy decisions are made—along with other stakeholders, from workers to universities.

For corporations that choose to be off the "American" team—to treat the United States at arm's length—access to American policymaking must be closed, and we should not promote or protect their interests, either here or abroad. This should be the criteria when considering which companies' executives should sit on advisory committees to the federal government, which companies should receive support from the State Department when competing for contracts overseas, which companies should have access to favorable financing terms from government agencies like the Overseas Private Investment Corporation (OPIC), and who should be consulted informally by key policymakers on questions of U.S. international economic strategy. By this logic, companies that are not headquartered in the United States but are committed to investing in this country and creating good jobs here should have a seat at the table, even if they are not technically "American" enterprises.

In the Globalism Era some firms will choose to be truly global—to owe no nation loyalty—and some will choose to reaffirm their identity as American enterprises. Both can contribute to the U.S. economy, but only one kind can rightly claim to be an "American" enterprise, and only one kind should be treated as part of our national community.

Sources

Balk, Alfred. 1990. *The myth of American eclipse: The new global era*. Rutgers, NJ: Transaction Publishers, 1990.

Lazonick, W. 2014. Profits without prosperity. *Harvard Business Review*, September 2014.

Prestowitz, C. 2010. *The betrayal of American prosperity*. New York: Free Press, 2010.

Tankersley, J. 2013. Wonkblog. *Washington Post*, July 17, 2013.

The Economist 2013. Welcome home. January 19, 2013.

Conclusion

THROUGHOUT ITS HISTORY, the United States has embraced capitalism as its fundamental economic system. In the words of economist Joseph Schumpeter, whose analysis of capitalism underlies the organization and themes of this book:

> The opening up of new markets, foreign or domestic, and the organizational development from the craft shop to such concerns as U.S. Steel illustrate the same process of industrial mutation—if I may use that biological term— that incessantly revolutionizes the economic structure from within, incessantly destroying the old one, incessantly creating a new one. This process of Creative Destruction is the essential fact about capitalism (1950, 83).

"Creative destruction," Schumpeter believed, was driven by continual innovation, increased efficiency, and expanding markets. It was powerful, but also disruptive. In the preceding pages, it is illustrated in the breaking of a British trade monopoly along the Red River; the creation and distribution of a practical sewing machine; the partnership of science and marketing that made Tupperware a success; and the development of genetically engineered soybeans, among many other episodes.

Capitalism is a powerful idea, but not an autonomous process. It exists only through the lives and activities of individuals, companies, organizations, and communities. Some who have led changes in American business history and appear in these pages are well known: Thomas Edison, Andrew Carnegie, Ruth Handler, and Sam Walton. Others noted may have surprised you: slave trader John Armfield, doll maker Portia Sperry, Annapolis deejay Charles W. "Hoppy" Adams, and supermarket sushi pioneer Ryuji Ishii. An important feature of capitalism is that it has taken many different forms in the lives of American business pioneers.

Even more important is recognizing that capitalism always exists in a social context, which determines how it affects communities and nations. Throughout U.S. history, it has been moderated by democratic forces and a shared social commitment to "common good." Sometimes these moderating forces have been the actions of federal and local governments, as in the abolition of slavery, the enactment of the New Deal, and the implementation of environmental regulations. Other times they have arisen from the work of nonprofit organizations, professional associations, or decisions by individuals to "give back" to a nation that fostered their success. Increasingly, the marketplace itself is the location of much change as consumers uneasy about consequences of efficiency and security have driven corporations to modify their behavior. As nations become ever more interdependent, however, the social context for American business is broadening. Thus, this history raises a concluding question that it can inform, but not answer: as capitalism becomes increasingly global, how can nations work together to moderate the changes it brings and improve not only America's, but also the world's common good?

DAVID K. ALLISON
Associate Director for Curatorial Affairs,
National Museum of American History

The United States $100 bill, 2013. Innovative security features made it easy for merchants to validate the bill and difficult for criminals to counterfeit it.

Reference List

Front Matter

Fitzpatrick, J. C., ed. 1938. *The writings of George Washington from the original manuscript sources 1745–1799*, vol. 28 (December 5, 1784–August 30, 1786), 1938.

The Merchant Era: 1770–1850s

Annual Report of the American Historical Association. 1892. Washington, D.C.: Smithsonian Institution, 1892.

Deyle, S. 2004. The domestic slave trade in America. *The Chattel Principle: Internal Slave Trades in the Americas*. Ed. Walter Johnson. New Haven: Yale University Press, 2004.

Deyle, S. 2005. *Carry me back: The domestic slave trade in American life*. New York: Oxford University Press, 2005.

Elizabeth Latimer to John Latimer. 1833. *John Latimer papers*. Philadelphia: Historical Society of Pennsylvania, Manuscript Division, June 1, 1833.

Gilman, R. R., C. Gilman, and D. M. Stulz. 1979. *The Red River trails, oxcart routes between St. Paul and the Selkirk settlement, 1820–1870*. St. Paul: Minnesota Historical Society, 1979.

Johnson, W. 2013. *River of dark dreams: Slavery and empire in the cotton kingdom*. Cambridge: Belknap Press of Harvard University Press, 2013.

Kerr, P. F. 1996. *Letters from China: The Canton-Boston correspondence of Robert Bennet Forbes, 1838–1840*. Mystic: Mystic Seaport Museum, Inc., 1996.

Morehouse, G. P. 1908. The Kansa or Kaw Indians and their history. *Collections of the Kansas State Historical Society* 10:345.

Peale, R. 1816. *Federal Gazette and Daily Advertiser*, June 13, 1816.

Petition of Is-tata Sin and others to Abraham Lincoln. 1863. Box 442, Record Group 75, Entry 79, Records of the Bureau of Indian Affairs, General Records, 1824–1907, Letters Received, 1824–1880. National Archives, Washington, D.C., July 17, 1863.

Ramsay, W. T. 1999. *The Ramsays: First family of Alexandria, Virginia*. Alexandria: W.T. Ramsay, 1999.

Ridge, M. 1962. *Cheer for the West*. St. Paul: Minnesota Historical Society, 1962.

Salem Gazette, November 16, 1784, 4.

Salem Mercury, February 26, 1788, 3.

Tadman, M. 1989. *Speculators and slaves: Masters, traders, and slaves in the old South*. Madison: University of Wisconsin Press, 1989.

Times Picayune, New Orleans, April 14, 1852, C3.

Vaiden, M. F. 1858. Policy on William, Virginia Historical Society, 1858.

William Ramsay's Will, Archives Center, National Museum of American History, Smithsonian Institution, Washington, D.C., 1852.

Winslow, S. N. 1864. *Biographies of successful Philadelphia merchants*. Philadelphia: James K. Simon, 1864.

The Corporate Era: 1860s–1930s

Calder, L. 1999. *Financing the American dream: A cultural history of consumer credit*. Princeton: Princeton University Press, 1999.

Doyle, C. 2013. John E. Powers (1834–1919). *A dictionary of marketing*. Oxford: Oxford University Press, 2013.

Foner, E., and J. Garraty, eds. 1991. *The reader's companion to American history*. Boston: Houghton Mifflin Company, 1991.

Ford, H. and S. Crowther. 1922. *My life and work*. Garden City: Garden City Publishing Co., Inc., 1922.

George Mason University. Transcription of President Franklin Delano Roosevelt's 1st inaugural address, 1933. http://historymatters.gmu.edu/d/5057/.

Nasaw, D. 2006. *Andrew Carnegie*. New York: Penguin Press, 2006.

New England Historical Society. Thomas Edison invents something no one wants. https://www.newenglandhistoricalsociety.com/thomas-edison-invests-something-one-wants/.

Rubenstein, H. 1989. Symbols and images of American labor: Badges of pride. *Labor's Heritage*, 1:2, April 1989.

Rutgers University. The Thomas Edison papers. http://edison.rutgers.edu/.

The Consumer Era: 1940s–1970s

Botkin, J. W., D. Dimancescu, and R. Stata. 1984. *The innovators: Rediscovering America's creative energy*. New York: Harper & Row, 1984.

DDB. Why Bernbach matters. http://www.ddb.com/ BillBernbachSaid/why_bernbach_matters/deep-influence.

Drucker, P. F. 1974. *Management: Tasks, responsibilities, practices*. New York: Harper & Row, 1974.

Flink, J. J. 1988. *The automobile age*. Cambridge: MIT Press, 1988.

Hamil, K. 1953. Working wife, $96.30 a week. *Fortune* April 1953, 158–168.

Hartmann, S.M. 1994. Women's employment and the domestic ideal. *Not June Cleaver: Women and Gender in Postwar America, 1945–1960*. Ed. Joanne Meyerowitz. Philadelphia: Temple University Press, 1994.

Hyman, L. 2011. Ending discrimination, legitimating debt: The political economy of race, gender and credit access in the 1960s and 1970s. *Enterprise and Society* 12:200–232.

IBM. Archives. http://www-03.ibm.com/ibm/history.

Jones, C. R. Papers. Archives Center, National Museum of American History, Smithsonian Institution.

Kaputa, C. 2009. *The female brand: Using the female mindset to succeed in business*. Boston: Davies-Black, 2009.

King, Jr. M.L. Where do we go from here?, delivered at the 11th Annual SCLC Convention, Atlanta, GA, August 16, 1967. The King Research and Education Institute, Stanford University, online speeches. http://mlk-kpp01.standford.edu.

Lubar, S. 1993. *Infoculture: The Smithsonian book of information age inventions*. Washington, D.C.: Smithsonian Press, 1993.

McCraw. T. K. 2009. *American business since 1920: How it worked*, 2nd ed. Wheeling: Harlan Davidson, Inc., 2009.

Mills, C. W. 1951. *White collar: The American middle classes*. New York: Oxford University Press, 1951.

National Equal Pay Task Force, 2013. Fifty years after the equal pay act: Assessing the past, taking stock of the future. The White House, June 2013.

New York Times 1958. Claire McCardell, designer, is dead. March 23, 1958.

Pursell, C. 2007. *Technology and the postwar era: A history*. New York: Columbia University Press, 2007.

Ryan, A., G. Trumbull, and P. Tufano, 2011. A brief postwar history of consumer finance. *Business History Review* 85:461–98.

Sanders, C. H. 2012. *The autobiography of the original celebrity chef*. Louisville: KFC Corporation, 2012.

Weems, R. 1998. *Desegregating the dollar: African American consumerism in the twentieth century*. New York: New York University Press, 1998.

Zumello, C. 2011. The everything card and consumer credit in the United States in the 1960s. *Business History Review* 85:555.

The Global Era: 1980s–Present

Charles, D. 2001. *Lords of the harvest: Biotech, big money, and the future of food*. Cambridge: Perseus Publishing, 2001.

Diaz, Z. trans. Christine Renee Miranda. La chiquita grows day by day. *Washington Hispanic*. http://www.washingtonhispanic.com/nota16144.html.

Doyle, B., e-mail message to Peter Liebhold, May 19, 2014.

Intel. Tech innovation: The pursuit of Moore's law, October 2011. http://newsroom.intel.com/community/news/blog/2011/10.

Konformist. A Buzzflash interview with Robert Reich, June 17, 2004. http://konformist.com/archives/2004/06-2004/rreich.txt.

MedicineNet. Paul Lewis: letter to the editor, *New York Times*, June 16, 1992. http://www.medterms.com/script/main/art.asp?articlekey=24845.

Reisman, G. 1998. *Capitalism: A treatise on economics*. Ottawa, IL: Jameson Books, 1998.

Shah, U., e-mail message to Peter Liebhold, June 3, 2014.

The Humane Society of the United States 1999. Flyer: The WTO: Have we traded away our right to protect animals. Division of Political History, National Museum of American History, 1999.

Conclusion

Schumpeter, J. A. 1950. *Capitalism, socialism and democracy*, 3rd ed. New York: Harper & Brothers, 1950.

Acknowledgments

PREPARING THE EXHIBITION *American Enterprise* **and this companion publication has been a journey of discovery and collaboration. The breadth of the work, both in time periods covered and subject matter** explored, meant that we as authors needed the insight and counsel of many. We are grateful for the time and energy that so many people extended to us making this undertaking possible.

We are indebted to the scholars who shared generously of their knowledge and offered excellent advice: Joyce Appleby, William Becker, Rebecca Blank, Steven Currall, Jane Gerhard, Jordan Grant, Mary Eschelbach Hansen, Roger Horowitz, David Hounshell, Thomas F. Jackson, Katie Knowles, Steven Lubar, Allison Marsh, Ann Smart Martin, Dan Mattausch, Vicki Mayer, David McDonald, Stephen Mims, Pietra Rivoli, Julia Sandy-Bailey, Susan Smulyan, Susan Strasser, Juliet E.K. Walker, and David Weinstein.

Numerous consultants offered their personal experiences and insights on American business: Ted Craver, Steven Fink, Jim McNulty, Michael Milken, Guillermo Nicolas, Patrick Parkinson, Carole and Gordon Segal, Twig Strickler, and Patricia Wanniski.

We relied heavily on the help and cooperation of many other museums and libraries. We would especially like to thank our colleagues Karen Kosanovich at Bureau of Labor Statistics, Ellen Terrell and Jennifer Harbster at the Library of Congress, and Paul Friedman at the New York Public Library.

We are grateful for the assistance from the following individuals and their organizations for their contributions in our search for photographs, documents, objects, and for answering our numerous queries: Barbara Flanaghan, Barbara Kaufman, and Bob McDonald at 3M; Marsha Appel at 4A's; Ryuji Ishii and Advanced Fresh Concepts; Scott Paul at the Alliance for American Manufacturing; Stephanie O'Keefe and the American Bankers Association; Dan Durheim, Pettus Read, and Mace Thornton at the American Farm Bureau; Chaya Brassarie and chef Shigefumi Tachibe; Jamal Booker, Justine Fletcher, Phil Mooney, and Ted Ryan at the Coca-Cola Company; Jonathan Kemper and Richard Moore at Commerce Bank; Jorge Rey at the U.S. Department of State; Jennifer Goldston, Michelle Gowdy, and Dean Oestreich at DuPont Pioneer; Samantha Cabaluna and Myra and Drew Goodman at Earthbound Farm; James Putnam at Farm Credit East; Brian Doyle, Jodelle French, Yogeeswaran Ganesan, Mario Paniccia, Uday Shah, and Rob Willoner at Intel; Sue Watson at Kraft Foods; Gail Broadright, Brad Figel, Sharon Heffelfinger, Lynne Rolland, Harold Schmitz, Howard-Yana Shapiro, and Rodney Snyder at Mars, Incorporated; Robert Fraley, Roy Fuchs, Brian Martinell, Heather McClurg, and Tami Craig Schilling at Monsanto Corporation; Janet Linde and Steve Wheeler at the New York Stock Exchange; Tony Jahn and Glyn Northington at Target Corporation; Greg Anderegg, Terri Boesel, and Jam Stewart at SC Johnson; Roy Bardole, Philip Bradshaw, Neil Caskey, and Sharon Covert at the United Soybean Board; Anne Effland, Susan Fugate, Perry Ma, and Lynn Stanko at the U.S. Department of Agriculture; Alan Dranow and Nick Graves at Walmart; and Marianne Babel at Wells Fargo.

A generous gift from Mr. and Mrs. Jonathan Kemper (William T. Kemper Foundation) made it possible for us to enhance the visual quality and readability of the book. We benefitted greatly from the design expertise of Bill Anton and editing support of Evie Righter. We also appreciate the help of Andrew Serwer, who worked closely with us to identify and guide the guest essayists who joined us in this project.

For providing us with so many wonderful illustrations, we are indebted to the photographers and staff of Smithsonian Photographic Services, including Jaclyn Nash, Hugh Talman, Richard Strauss, and Harold Dorwin.

We are grateful for the talented editorial team at Smithsonian Books, Carolyn Gleason, Christina Wiginton, and Raychel Rapazza, for patiently guiding us through the publication process.

The cooperation of over a hundred Smithsonian staff, from conservators and registrars to curators and collections managers, enriched our efforts. We are especially indebted to: Nancy Bercaw, Doris Bowman, Eduardo Diaz, Janice Ellis, Jane Fortune, Paul Gardullo, Carolyn Gilman, Katharine Klein, Bonnie Campbell Lilienfeld, Konrad Ng, Sunae Evans Park, Bill Pretzer, Beth Richwine, Harry Rubenstein, Noriko Sanefuji, Wendy Shay, Gabrielle Tayac, Suzanne Thomassen-Krauss, Christopher Turner, Hal Wallace, Mallory Warner, Leslie Wilson, Tim Winkle, and Bill Yeingst.

Our heartfelt thanks to the many volunteers and interns who contributed to the development of this book, including: Ann Abney, Ailyn Alonso, Chris Fite, Brian Johnson, Lindsay Keating, Joan Krammer, Francesca Lo Galbo, Halle Mares, Shagun Raina, Rebecca Soules, and Susan Strange.

Illustration Credits

American Association of Advertising Agencies: p. **72**

AP Photo/Dave Pickoff: p. **144L**

Chicago History Museum: i08679, p. **83**; DN-0008625, p. **85**

Colonial Williamsburg Foundation: p. **30**

Compliments of Mars, Incorporated: p. **186**; p. **187**

Copyright © 2014 Intel Corporation: p. **156–157**; p. **189**; p. **190TL**; p. **190BL**

© Renee Farias: p. **184L**

Courtesy of Alexandra Whyte: p. **112**

Courtesy of the Boston Public Library: Leslie Jones Collection, p. **94L**

Courtesy of Cagle Cartoons, Inc.: p. **195**

Courtesy of Cortez Family: p. **131**

Courtesy of DDB Worldwide: p. **136**; p. **137R**

Courtesy of Foster Farms: p. **179**

Courtesy of GM Media Archives: p. **70L**

Courtesy of IBM Archives: p. **118T**; p. **119**

Courtesy of Jack Corn Collection, Vanderbilt University Special Collection: p. **114**

Courtesy of the Library of Virginia: p. **8**

Courtesy of Martin Agency on behalf of GEICO: p. **174**

Courtesy of Michael Milken: p. **166L**

Courtesy of the Rare Book & Manuscript Library, University of Illinois at Urbana-Champaign: p. **79**

Courtesy of Rhode Island Historical Society: James DeWolf, c. 1825, watercolor on ivory, RHi X4 213, p. **35**

Courtesy of Rockefeller Archive Center: p. **75**

Courtesy of the S. Gansean family: p. **181T**; p. **182**

Courtesy of the Sperry Family: p. **98L**

Courtesy of Wallace H. Coulter Foundation: p. **191**

Courtesy of Wells Fargo: p. **168**

CSG CIC Glasgow Museums Collection: p. **20R**

The Drucker Institute at Claremont Graduate University: p. **120L**

Library and Archives Canada: Acc. No. 1973-84-1, p. **28**

The Library Company of Philadelphia: p. **52**

Library of Congress: Geography and Map Division: p. **41L**; Prints and Photographs Division: LC-USZC2-4912, p. **55**; LC-USZ6-787, p. **64**; LC-USZCN4-122, p. **73**; National Child Labor Committee Collection, LC-DIG-nclc-01824, p. **81**; LC-USZ62-119495, p. **82L**; LC-USZ62-92431, p. **88**; LC-DIG-ppmsc-00816, p. **99**; LC-USZ62-17305, p. **101**

Minnesota Historical Society: p. **42**

Munro Leaf papers, Free Library of Philadelphia, Rare Book Department: p. **128**

Museum of Fine Arts, Boston: Henry H. and Zoe Oliver Sherman Fund, p. **2**; Gift of Joseph W. Revere, William B. Revere and Edward H. R. Revere, p. **23T**

National Archives and Records Administration: Bureau of Prisons, p. **93T**

National Museum of American History, Smithsonian Institution: SI Neg. JN2014-3879, Cat. No. 2014.3077.08, p. **6**; SI Neg. JN2014-3520, Cat. No. 1987.0910.0001, p. **9**; SI Neg. JN2014-3381, Cat. No. 006544, p. **10**; SI Neg. 91-819, p. **11T**; SI Neg. ET2014-41177, Cat. No. 1985.0460.129, p. **11B**; SI Neg. JN2014-3921, Cat. No. 1987.0127.01, p. **12**; SI Neg. JN2014-3626, Cat. No. 312108, p. **13L**; SI Neg. ET2014-12053, p. **13R**; SI Neg. ET2014-41015, Cat. No. 2008.0074.01, p. **17T**; SI Neg. JN2014-3411, Cat. No. 6724, p. **17B**; SI Neg. JN2014-3288, Cat. No. T13660.00A, p. **18**; SI Neg. JN2014-3606, Cat. No. 1984.0486.06, p. **19**; SI Neg. JN2014-3586, Cat. No. 388196, p. **20L**; SI Neg. JN2014-3622, Cat. No. 62.0549, p. **21**; SI Neg. JN2014-3610, Cat. No. 2014.0015.01, Cat. No. 2014.0015.04, p. **22BL**; SI Neg. JN2014-3387, Cat. No. 62.711, p. **22BRL**; SI Neg. ET2014-40555, Cat. No. 15095, p. **22BRR**; SI Neg. ET2012-13578, Cat. No. 60.2211, p. **23B**; SI Neg. JN2014-3518, Cat. No. 81995, p. **24T**; SI Neg. JN2014-3425, Cat. No. 294439.11, p. **25T**; SI Neg. JN2014-3435, Cat. No. 020055, p. **25B**; SI Neg. JN2014-3522, Cat. No. 1979.1263.00755, p. **26BL**; SI Neg. JN2014-3225, Cat. No. 15678, p. **26BR**; SI Neg. RWS2013-01080, Cat. No. 1979.0425.183, p. **28BFL**; SI Neg. JN2014-3612, Cat. No. 195102.000, p. **28BL**; SI Neg. JN2014-3400, Cat. No. 1988.0129.01, p. **28BR**; SI Neg. JN2014-3857, Cat. No. 180182, p. **29**; SI Neg. JN2014-3428, Cat. No. 1983.0853.01, p. **31R**; SI Neg. 227739.1835, Cat. No. 227739.1835, p. **33**; SI Neg. ET2011-09609, Cat. No. 2007.0093.01, p. **34**; SI Neg. JN2014-3572, 001252,

p. **43L**; SI Neg. JN2014-3519, Cat. No. 283645.1372,
p. **44T**; SI Neg. JN2014-3497, Cat. No. 61.8, p. **45R**;
SI Neg. ET2014-40645, Cat. No. 050039, p. **46TR**; SI Neg.
ET2014-40516, Cat. No. 033487, p. **46AB;** SI Neg.
ET2014-40588, Cat. No. 273137, p. **47**; SI Neg. JN2014-
3274, Cat. No. 2012.0106.01, p. **48**; SI Neg. JN2014-3402,
Cat. No. 057605B, p. **49T**; SI Neg. JN2014-3542, Cat. No.
332316, p. **49B**; SI Neg. JN2014-3230, Cat. No.
1995.0001.02, p. **51L**; SI Neg. JN2014-3624, Cat. No.
181797, p. **62BL**; SI Neg. JN2014-3625, Cat. No. 252616,
p. **62TL**; SI Neg. JN2014-3598, Cat. No. 311215, p. **62R**;
SI Neg. JN2014-3602, Cat. No. T06054.000, p. **66B**;
SI Neg. JN2014-3626, Cat. No. 312108, p. **67**; SI Neg.
JN2014-4186, Cat. No. 328538, p. **68L**; SI Neg. JN2014-
3248, Cat. No. 2013.0323.02, p. **68BR**; SI Neg. JN2014-
3477, Cat. No. 2013.0323.01, p. **68TR**; SI Neg. JN2014-
4618, Cat. No. 323505.01, p. **71L**; SI Neg. JN2014-3335,
Cat. No. 2014.0072.15, p. **71R**; SI Neg. JN2014-3618,
Cat. No. 252599, p. **76R**; SI Neg. JN2014-3514, Cat. No.
1989.0396.007, p. **80L**; SI Neg. JN2014-3511, Cat. No.
1989.0693.4470, p. **80R**; SI Neg. JN2014-3515, Cat. No.
1989.0693.2064, p. **82R**; SI Neg. JN2014-3558, Cat. No.
2014.0062.13, p. **86**; SI Neg. JN2014-3574, Cat. No.
2014.0062.05, p. **87L**; SI Neg. ET2014-40630, Cat. No.
2006.0098.1351, p. **87R**; SI Neg. ET2014-41534, Cat. No.
309201, p. **89B**; SI Neg. JN2014-3524, Cat. No.
1986.0873.55, p. **90L**; SI Neg. ET2014-1070, Cat. No.
66A09, p. **90R**; p. 90-91; p. **92**; SI Neg. JN2014-3490,
Cat. No. 69052M, p. **92–93**; SI Neg. JN2014-3233,
Cat. No. 2012.0027.02, Cat. No. 1979.0798.509, p. **94R**;
SI Neg. JN2014-4386, Cat. No. 1979.0798.509, p. **95T**;
SI Neg. JN2014-3629, Cat. No. 1980.0954.01-.09, p. **95B**;
SI Neg. AHB2008q10508, Cat. No. 83.69.7, p. **96**; SI Neg.
JN2014-3623, Cat. No. 1986.0671.01, p. **97**; SI Neg.
JN2014-3241, Cat. No. 1979.1263.00472, p. **98R**; SI Neg.
JN2014-4616, Cat. No. 323538.01, p. **100**; SI Neg.
JN2014-3483, Cat. No. 1989.0693.0252, p. **102T**; SI Neg.
JN2014-3481, Cat. No. 1989.0693.0302, p. **102C**; SI Neg.
JN2014-3486, Cat. No. 1979.0976.29, p. **102B**; SI Neg.
JN2014-3802, p. **110L**; SI Neg. JN2014-3619, p. **110R**;
SI Neg. ET2014-40491, Cat. No. 1980.0316.12, p. **113R**;
SI Neg. JN2014-3498, Cat. No. 2013.0129.02, p. **118B**;
SI Neg. JN2014-3404, Cat. No. 1995.0248.02, p. **120ATR**;
SI Neg. JN2014-3575, Cat. No. 2013.3049.02, p. **120ATC**;
SI Neg. JN2014-3282, p. **120CR**; SI Neg. ET2014-41384,
p. **120BR**; SI Neg. JN2014-3617, Cat. No. 2014.0120.01,
p. **124**; SI Neg. JN2014-3599, Cat. No. 1988.3107.05,
p. **126TL**; SI Neg. JN2014-3246, p. **126TR**; SI Neg.
JN2014-3509, Cat. No. 327729, p. **126B**; SI Neg. ET2014-

40406, Cat. No. 1980.0910.05, p. **129**; SI Neg. JN2014-
3426, p. **130**; SI Neg. ET2014-40935, Cat. No. 84.17.26,
p. **132L**; SI Neg. ET2014-40915, Cat. No. 72.66.36,
p. **132R**; SI Neg. ET2014-40932, Cat. No. 84.17.11,
p. **133B**; SI Neg. ET2012-11280, Cat. No. 75.55.32,
p. **133T**; SI Neg. RWS2014-01958, Cat. No. 75.55.20,
p. **134T**; SI Neg. RWS2014-01956, Cat. No. 75.55.18,
p. **134B**; SI Neg. RWS2014-01962, Cat. No. 75.55.16,
p. **135TL**; SI Neg. RWS2014-01954, Cat. No. 75.55.19,
p. **135BL**; SI Neg. RWS2014-01964, Cat. No. 75.55.17,
p. **135TR**; SI Neg RWS2014-01966, Cat. No. 75.55.15,
p. **135BR**; SI Neg ET2012-11279, Cat. No. 75.55.22,
p. **137L**; SI Neg. JN2014-3611, Cat. No. 316474.026,
p. **141R**; SI Neg. JN2014-3528, p. **142TL**; SI Neg. JN2014-
3529, p. **142TR**; SI Neg. JN2014-3596, Cat. No.
2002.0319.20, p. **150B**; SI Neg. JN2014-3567, Cat. No.
1988.0520.022, p. **159AT**; SI Neg. JN2014-3568, Cat. No.
1985.0106.386, p. **159AC**; SI Neg. JN2014-3569, Cat. No.
1989.0693.0650, p. **159AB**; SI Neg. ET2014-40411,
Cat. No. 2001.0072.01, p. **159B**; SI Neg. ET2014-40437,
Cat. No. 2004.0258.01, p. **160**; SI Neg. ET2014-40454,
Cat. No. 2012.0134.01, p. **161TL**; SI Neg. ET2014-40446,
Cat. No. 1999.0155.01, p. **161BL**; SI Neg. JN2014-3427,
Cat. No. 1995.0085.002, p. **162**; SI Neg. 2000-7799,
Cat. No. 2000.0158.01, Cat. No. 2000.0158.02, p. **163TL**;
SI Neg. JN2014-3492, Cat. No. 2000.0266.01, p. **163TR**;
SI Neg. JN2014-3580, Cat. No. 2014.0072.04, p. **164T**;
SI Neg. JN2014-3537, Cat. No. 2000.0266.02, p. **164B**;
SI Neg. JN2014-3573, Cat. No. 2014.0056.01, p. **165**;
SI Neg. JN2014-3576, Cat. No. 2013.0150.01, p. **166R**;
SI Neg. JN2014-3226, Cat. No. 1989.0402.19, p. **167**;
SI Neg. JN2014-3587, p. **169L**; SI Neg. JN2014-3934,
Cat. No. 2013.0219.01, p. **169R**; SI Neg. JN2014-3531
Cat. No. 2014.0111.01, p. **170**; SI Neg. JN2014-3505,
Cat. No. 2013.0271.01, p. **171**; SI Neg. ET2014-40435,
2013.0024.01, p. **175T**; SI Neg. JN2014-3545, Cat. No.
1993.0354.01, p. **175B**; SI Neg. ET2014-40413, Cat. No.
2011.0173.08, p. **176**; SI Neg. ET2014-40465, Cat. No.
2012.0166.01, p. **177**; SI Neg. ET2014-40412, Cat. No.
2013.0022.01, p. **178R**; SI Neg. JN2014-3578, Cat. No.
2011.0255.04, p. **180**; SI Neg. JN2014-3375, Cat. No.
2014.0038.02, p. **181B**; SI Neg. JN2014-3689, p. **183L**;
SI Neg. JN2014-3571, Cat. No. 2000.0198.816, p. **183TR**
left corner; SI Neg. JN2014-3570, Cat. No. 1999.0322.53,
p. **183BR** left corner; SI Neg. JN2014-3592, Cat. No.
1999.0322.53, p. **183TR** right corner; SI Neg. JN2014-
3591, Cat. No. 2000.0198.40, p. **183BR** right corner;
SI Neg. JN2014-3970, Cat. No. 2007.0218.03, p. **184R**;
SI Neg. JN2014-3969, Cat. No. 2014.0013.01, p. **185**;

SI Neg. JN2014-3688, Cat. No. 2014.0137.01, p. **188**;
SI Neg. JN2014-3275, Cat. No. 2003.0231.17, p. **190R**;
SI Neg. RWS2013-01555, Cat. No. 2013.0041.01, p. **192**;
SI Neg. JN2014-3555, Cat. No. 2013.3116.01, p. **194**;
SI Neg. JN2014-3609, Cat. No. 2013.3097.02, p. **196**;
SI Neg. ET2014-41110, Cat. No. 2012.0245.03, p. **197**

Archives Center:

Allen Balcom Du Mont Collection, 1929-1965, Archives Center: SI Neg. 88-628, p. **125**

American Petroleum Institute Photograph and Film Collection, Archives Center: SI Neg. AC0711-0000040, p. **60–61**, **74**

Archives Center: SI Neg. JN2014-3526, Cat. No. 229563.007A, p. 2**4R**

Arthur Bernie Wood Papers, Archives Center: SI Neg. AC0962-0000017, p. **121**; SI Neg. AC0962-0000037-01, p. **123**

Brownie Wise Papers, Archives Center: SI Neg. AC0509-0000005, p. **108–109**, p. **142B**; SI Neg. AC0509-0000018, p. **144R**

Carolyn R. Jones Collection, Archives Center: SI Neg. AC0552-0000181-01, p. **147L**; SI Neg. AC0552-0000188, p. **147R**

Estelle Ellis Collection, Archives Center: SI Neg. AC0423-0000025-01, p. **138**; SI Neg. AC0423-0000019, p. **139**; SI Neg. AC0423-0000003, p. **140**

Frank and Lillian Gilbreth Collection, Archives Center: SI Neg. AC0803-0000015, p. **69**; SI Neg. 89.14251, p. **70R**

Fred/Alan MTV Network Advertising Collection, Archives Center: SI Neg. AC0453-0000019, p. **161R**

N. W. Ayer Advertising Agency Records, Archives Center: SI Neg. AC0059-0000012, p. **77**; SI Neg. AC0059-0000143, p. **78**

On loan from Google, Inc.: SI Neg. JN2015-5049, p. **158**

Ramsay Family Papers, Archives Center: SI Neg. JN2014-3604, p. **22T**; SI Neg. AC0088-0000002-4, p. **32**

Scurlock Collection, Archives Center: SI Neg. AC0618.004, p. **146**

WANN Radio Station Records, Archives Center: SI Neg. AC0800-0000001, p. **149**; SI Neg. 03080001, p. **150T**

Warshaw Collection of Business Americana, Archives Center: SI Neg. 85-14366, p. **66T**

Courtesy of Patricia Koyce Wanniski: SI Neg. JN2013-1736, p. **193**

From the Collections of Larry Bird: p. **84**

Lent by Dan Mattausch and Nancy Mattausch: SI Neg. JN2014-3390, p. **51R**; SI Neg. JN2014-3391, p. **54**

National Archives and Records Administration: SI Neg. RWS2013-01248, p. **38T**; SI Neg. RWS2013-01249, p. **38B**; SI Neg. RWS2013-01250, p. **39T**; SI Neg. RWS2013-01251, p. **39B**; SI Neg. RWS2013-01253, p. **40**

Superman is ™ and © DC Comics: SI Neg. JN2014-3588, p. **117**

National Museum of American Indian, Smithsonian Institution: SI Neg. JN2014-3620, p. **37R**

National Portrait Gallery, Smithsonian Institution: Gift of Susan Mary Alsop, p. **27**; p. **41R**; Gift of Margaret Carnegie Miller, p. **76L**; Gift of Time magazine, p. **141L**; Gift of Estrellita Karsh in memory of Yousuf Karsh, p. **172**

Nebraska State Historical Society: RG2608-1784, p. **89T**

New York Public Library: p. **14–15**, p. **46L**

Peabody Essex Museum: E79708, p. **44B**; Gift of Rebecca B. Chase, Ann B Mathias, Charles E. Bradford, 1990, M23228, p. **45L**

Philadelphia Museum of Art: The Collection of Edgar William and Bernice Chrysler Garbisch, 1965, p. **16**, p. **50**

Photo Courtesy of Earthbound Farm: p. **178L**

Photo courtesy of USDA Natural Resources Conservation Service: p. **103**

Photo by Peter Yates: p. **163B**

Photo by Sovfoto/UIG via Getty Images: p. **111**

Photograph courtesy of Monsanto Company: p. **173**

Smithsonian American Art Museum: Gift of Mrs. Joseph Harrison, Jr., p. **36**, p. **37L**

Smithsonian Libraries: SI Neg. JN2014-4112, Cat. No. 63.633PA, p. **24B**

Tennessee State Library and Archives in Nashville: p. **31L**

UCLA: *Los Angeles Times* Photographic Archive, Department of Special Colections, Charles E. Young Research Library, p. **143**

University of Maryland Libraries: Special Collections & University Archives, The George Meany Memorial AFL-CIO Archive, 1st Convention of the AFL-CIO, December 5, 1955, copyright AFL-CIO, Photo by Frank Alexander, Merkle Press, p. **113L**

University of Texas at San Antonio Libraries Special Collections: p. **43R**

Used with permission of Intel Corporation: p. **116L**, p. **116R**

Virginia Historical Society: p. **33B**

The Walters Art Museum, Baltimore: p. **26T**; p. **53**

Index